Chromosomal Diversity
in Insects

The Authors

Dr. Sathe Tukaram Vithalrao [M.Sc., Ph.D., Sangit Vishard, IBT (Seri.), F.I.S.E.C., F.S.E.Sc., F.S.L.Sc., F.I.C.C.B., F.S.S.I., FHAs] is presently working as Professor and former Head, Department of Zoology, Shivaji University, Kolhapur. He has teaching experience of 29 years in Entomology at University PG department and 15 years in Agrochemicals and Pest Management. He has written 33 books and published 255 research papers in national and international journals of repute. He guided 25 Ph.D. students and completed 6 major research projects (from CSIR, DST, DBT and UGC). He visited Canada (1988), Japan (1988), Thailand (2002, 2004), Spain (2005), France (2005), South Korea (2006) and Nepal (2007) etc. for academic work. He is member of editorial board of eleven prestigious journals. He delivered 35 talks through All India Radio and international conferences and involved in Doordarshan, S.T.V. and B.T.V. programmes on useful and harmful insects. He published more than 35 popular articles in daily newspapers on insects and sericulture. He got several prestigious awards like "Environmentalists of the Year-2003", "Bharat Jyoti", "Jewel of India", "International Gold Star", "Eminent Citizen of India", "Education Acumen", "Best Educationist", "Eminent Scientist of the Year-2008", "Lifetime Education Achievement", "Lifetime Achievement in Zoology (Insect Taxonomy)-2009", Education Leadership-2011, Asia Pacific International Award-2012, Global Education Leadership Award-2013, etc. He is also working as Research and Recognition (RR) Committee member for Pune University, Pune; North Maharashtra University, Jalgaon; Shivaji University, Kolhapur and DBA Marathwada University, Aurangabad. He has been awarded several fellowships from different scientific and academic societies. He is Chairman of Maharashtra District Environmental Centre of NESA.

Dr. Satywan Subarao Patil (M.Sc., Ph.D.) is presently working as Assistant Professor in Department of Zoology, ACS College, Palus, Dist. Sangli. His specialization is cell biology and have twenty one year teaching experience. He has published number of research papers in national and international journals of repute. He has participated/delivered lectures at various conferences/symposium etc. He was BOS Member of Zoology, Shivaji University, Kolhapur. He is chairman of Environmental Centre, Sangli District.

Chromosomal Diversity in Insects

— Authors —

T.V. Sathe

Professor & Former Head
Department of Zoology
Shivaji University
Kolhapur – 416 004, M.S.

S.S. Patil

Associate Professor
Department of Zoology
A.C.S College, Palus
Sangli, M.S.

2015

Daya Publishing House®

A Division of

Astral International Pvt. Ltd.

New Delhi – 110 002

Published by   : **Daya Publishing House®**
                            A Division of
                            Astral International Pvt. Ltd.
                            – ISO 9001:2008 Certified Company –
                            4760-61/23, Ansari Road, Darya Ganj
                            New Delhi-110 002
                            Ph. 011-43549197, 23278134
                            E-mail: info@astralint.com
                            Website: www.astralint.com

Laser Typesetting   : **Classic Computer Services**, Delhi - 110 035

Printed at   : **Thomson Press India Limited**

PRINTED IN INDIA

Preface

Chromosomes are hereditary material and have very important role in speciation and identification of closely related species. Chromosomal diversity and their role in speciation and organic evolution has been discussed in different insects including beetles, mosquitoes, aphids, leafhoppers, grasshoppers etc. The chromosomes are basis for the genetic work. Hence, chromosomes have been reported in 8 species of Coccinellids. Simultaneously, morphological diversity have also been reported and correlated with chromosomes from Sangli district of Maharashtra, India. In all 19 species have been reported. Out of which 11 species are found new to the science. Various concepts of chromosomal diversity, genetics, evolution, speciation and insect pest management discussed in this book will be helpful and stimulatory to students, teachers, and scientists. We thank Shivaji University authority for providing facilities for this work. We also thank to Madhuri Sathe and Dr. Nishad Sathe for their help in completion of this book in various ways.

T.V. Sathe

S.S. Patil

Contents

	Preface	*v*
1.	General Introduction	1
2.	Review of Literature	9
3.	What is Chromosome?	30
4.	Collection, Preservation and Methodology	51
5.	Morphological Diversity of Coccinellids	62
6.	Chromosomal Diversity of Coccinellids	91
7.	Concepts of Chromosomal Diversity	110
	Bibliography	145
	Index	159

1

General Introduction

In recent years no much progress is seen in crop diversity and productivity in India (Figure 1). In fact, high yielding verities of crops need substantial quality of chemical fertilizers, pesticides and water, without which they cannot survive and produce higher yield. India has achieved a considerable degree of food security. But Indian agriculture is now under threat due to depletion of groundwater resources. The diminishing range of yield and agricultural technologies are slowing down the growth and food production of our country.

Forests are wealth for any country in the World. The forest is an ecological system of mainly three dominated vegetative associations which provides timber, fuel, wood, fodder and fiber grasses and non wood forest products and support industrial and commercial activities and related to ecological balance and life support system (food production, health and development) of man. Out of 329 million hectares of geographical area of our country, actual forest cover exists only 76.52 million hectares (Mohan Kumar, 2006). The actual recorded deforested and/or degraded forest area is much greater. Out of 413 districts of our country, 150 districts had

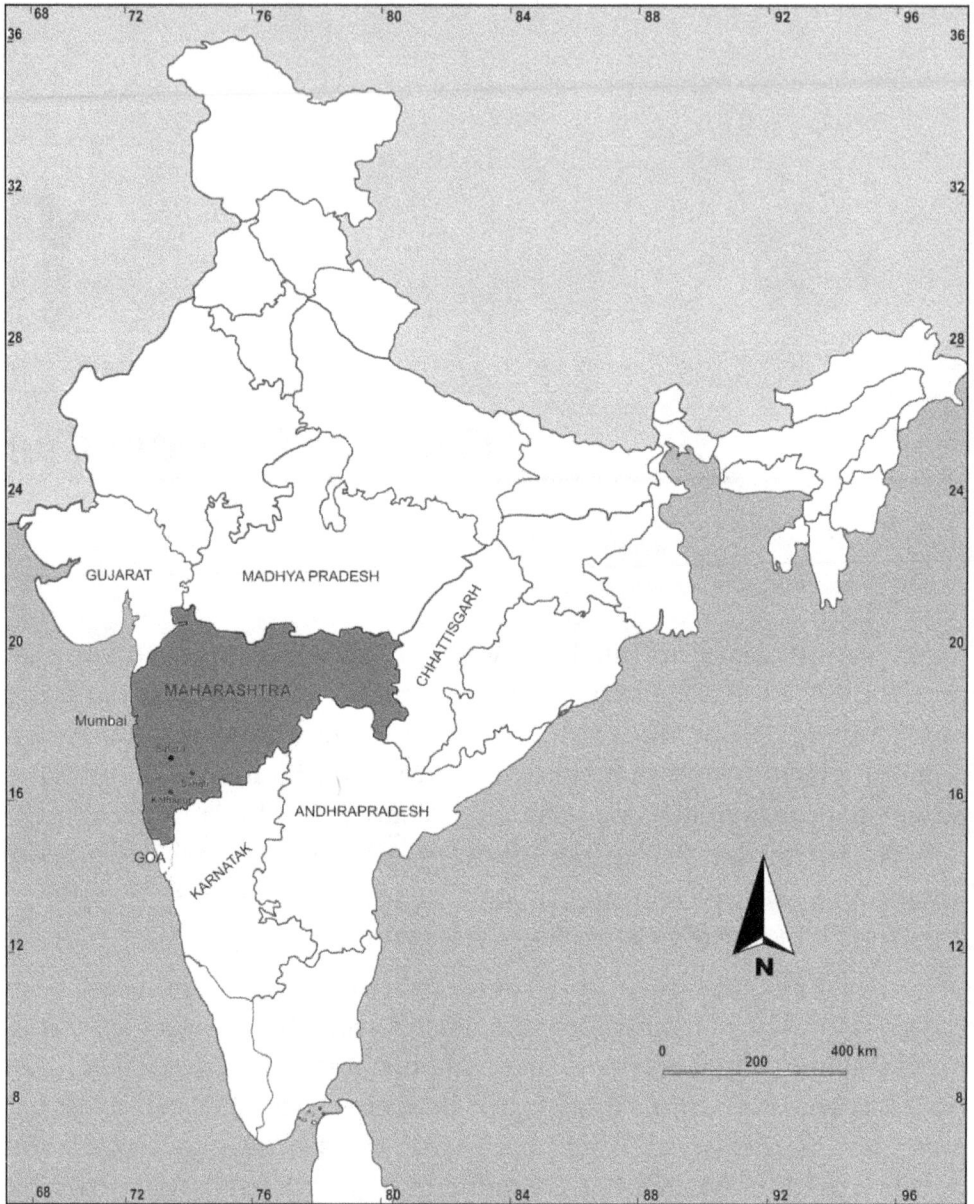

Figure 1: Map of India Showing Maharashtra and Study Area.

forest cover of more than 33 per cent, 52 districts are between 19 – 33 per cent and 226 districts have less that 19 per cent forest cover.

Indian forests are of four major groups, namely tropical, subtropical, temperate and alpine. These major groups are further divided in to 16 types which refer to tropical dry deciduous, tropical moist deciduous, tropical wet evergreen, subtropical pine, subalpine, Himalayan moist temperate, tropical semi evergreen, montane wet temperate, littoral and swamp, subtropical broad leaved hill, subtropical dry evergreen, Himalayan dry temperate, and tropical dry evergreen. The rate of losing forest cover in India is 144, 000 hectares per year, however, it was slowed down to 24,533 hectares during the years 1980 and 1995. India shows following very important forests based on species richness namely, Himalayan region, North eastern region and Western Ghats.

According to UNEP (1995) biodiversity is the variability among living organisms from all sources including terrestrial, marine and other aquatic ecosystems and the ecological complexes of which they are a part, this includes diversity within species, between species and of ecosystem. Diversity can also be defined as the species richness of plants, animals and micro-organisms, including their genetic makeup and the communities they form. In general, biodiversity has three components namely, genetic, organism and ecological with their own hierarchical importance, interlinked at any one level and have impact on other level of hierarchy and have continuity among the different component from genes to ecosystems. Biodiversity has ecological, intrinsic, productive and consumptive use values. Today's biodiversity is a product of over 3.5 billion years of evolution, involving speciation, migration, extinction and human influence. However, species are being lost much faster than the natural rate of extinction. A tremendous loss of genetic diversity has been noticed during the last hundred years. Many species are on the verge of extinction. Hence, Indian biodiversity is at a very high risk of extinction due to high endemism. Therefore, protection India's biodiversity is the need of the day and global concern.

According to Joseph (2004) ecological diversity is the diversity of ecosystems/habitats/biomes/bioregions/landscapes/communities that do not really exist as discrete units but are part of much wider system. Species diversity is the basic and simplest level biodiversity and it is normally considered as total number of different species in a given area known as 'species richness'. It is also sometime measured as the number different species and their relative abundance or the number of different trophic levels. The number of techniques has been developed to measure species richness, known as Indices, based on different assessments (Krebs, 1989). A diversity index is, therefore, a numerical expression of the amount of biodiversity observed within a sample of organisms. 'Biodiversity indexing' is different from biodiversity index. The former is the process of correctly identifying organisms, mapping their distributions to understand their ecological position and process of evolution, vouchering the identified organisms and integrating this information into international accessible data bases. Biodiversity can be indexed at different levels namely, landscapes, ecosystems, communities, species varieties and populations. Worldwide scientists have identified and described 17,00,000 species of micro-organisms, plants and animals. During the 25 years, 20,000 species have been identified and named per year. It is however estimated that there can be between 10,000,000 and 13,000,000 species on earth of which insects alone amount to 8, 000, 000 species. Insects, as we are aware, are important in global agricultural economy; micro-organisms, like insects, are important determinants of agricultural productivity. It is estimated that there can be 400,000 species of viruses, 400,000 species of bacteria and 150,000 species of fungi. Of the estimated total of 226,000 species of living plants, humans exploit 3000 as food. The tropical regions are generally richer in biodiversity than the temperate and arctic regions of the world. However, tropical rain forests and coral reefs are the best sources of biodiversity.

Joseph (2004) says that genetic diversity is the finer level and it refers to the variation among the individuals and between populations within a species, as measured by the variation in genes of a particular species, subspecies or breed/strain. Genetic variability within a population can be measured as the number (and percentage) of genes in the population that are polymorphic, the number of alleles for each polymorphic gene and the number (and percentage) of genes per individual that are polymorphic. Genetic diversity is the basic currency of evolution and adaptation and is affected by cytogenetic processes such as mutation and recombination, population/evolution phenomena like immigration, emigration, selection etc. and anthropogenic activities like habitat destruction, pollution,etc.

Rich biodiversity is an indicator of the health of a particular habit, bio-geographic area and its potential to sustain life. Loss of biodiversity is irreversible. The current rates of species extinction are 1000 to 10,000 times higher than the background rate inferred from fossil record. Today we seem to be losing 2 – 5 species per hour from tropical forests alone. Factors leading to the loss of biodiversity are both direct and indirect factors. These include habitat loss, invasion by introduced species, pollution, global climate change, over exploitation of species and certain agricultural and forestry practices. For most wild species, little information is available on the extent of genetically distinct populations and loss of genetic diversity. Reduction in population size will usually result in loss of genetic diversity and therefore adaptability of the species. The rich genetic variability in this biosphere should be conserved by all means to ensure sustainability of the delicate ecological balance.

India (Figure 1) is a vast country with a rich diversity of biotic resources and is ranked one of the 12 – mega diversity counties in the tropics. This rich biodiversity in India is largely due to her varied physical environment in terms of latitude, longitude, altitude, geology and climate (Rojers, 1991). India harbours two important global 'hot spots', Western Ghats and Eastern Himalaya. The 'hot spot biodiversity' concept has been introduced initially by Mayers (1988). He identified 10 regions of hot spots that are characterized by high concentration of endemic species and are experiencing rapid rate of extinction or loss. Hot spots are invariably associated with 'endemic biodiversity'. The endemic taxa may occur in a small – restricted area, an island, a peninsular, a bio-geographic area or a mountain top. According to Cherian (2000) several factors are responsible for the development and evolution of biodiversity of hot spots. These are impacts of transform faulting produced by the geological process, impact of mantle plumes, latitudinal diversity gradient, impact of solar energy, survival in refugial pockets through the ice-ages and impact of changes of topography, environment and drainage systems. On the basis of evolution or the trends of evolution and biodiversity of 18 hot spots recognized by Myers and the 24 by Mittermeier *et al.* (1998), it is possible to divide them in to four groups (1) hot spots falling within the zone of direct impact of the colliding plates. (2) Hot spots which fall just outside the zone of direct impact of the colliding plates. (3) Hot spots which have in directly between influenced the plates in the recent past and (4) Hot spots, the biodiversity of which partly originated in unison during the remote past (Cherian, 2000).

Biodiversity conservation of endemic centers and other geographical areas of interests is on the national agenda for Nations/States who are signatories of the Convention on biodiversity, which came into force on 29th December, 1993. According to Katzman and Cale (1990) man-made extinction of species and habitat degradation and loss are the order of the day. Species depletion is acute in tropical forests deforesting 100,000 – 200,000 sq km per annum. Against this background of habitat loss and degradation and the resultant species extinction, it is necessary to develop strategies for the protection of our ecosystems, species and genes. Above facts have relation with the human population pressures and the developmental needs of tomorrows. For conservation and protection of diversity *in situ* and *ex situ* conservation strategies should be supplemented as holistic/adoptable strategies. Lady bird beetles are natural living resource of insect pest control hence; they should be conserved, protected and utilized in pest control programmes.

India has very impressive agro-biodiversity hence ranks seventh in the world in number of species contributed to agriculture and animal husbandry. Agriculture provides the livelihood for the 3/4 of its population. Kumar (1990) estimated that by 2020 India will need 249 million tons of food grains. Therefore, yield of crop should be increased by 22 to 41 per cent in cereals, and 110 per cent in pulses is challenge to both agronomists and breeders in future because, insects cause enormous damage to agricultural crops, whenever hybrid crops are taken, insects would be therefore affecting the crop yields. Indiscriminate use of pesticides on crop and soil along with organic wastes, including domestic and industrial sewage, sludge and food processing etc. and causing serious environmental problems and pollution (Mukhopadhay *et al.*, 2005). Therefore, eco-friendly pest management has no option today.

Biological pest control is eco-friendly, pollution less and safety to humans. Lady bird beetles are biocontrol agents of several insect pests and are fully utilized in pest management (Sathe, 2004). Hence, any advance knowledge on Lady bird beetles specially, on chromosomal diversity will add great relevance in mass utility of these species in eco-friendly pest control techniques.

Very large number of insect species have been recorded from agro and forest ecosystems in India. About 67, 000 species of insect have been described (ZSI, 1983).The most dominant group of insects is Coleoptera which accounts 57 per cent followed by Lepidoptera, constituting 20 per cent (Nair *et al.*, 1981). From Indian forests about 16,000 species have been recorded. The present experimental work is largely related to Sangli District (Figure 2) and Lady bird beetles (Plates 1 and 2) since Sangli district is agriculturally forward district of Maharashtra and Lady bird beetles are very potential bio-control agents of insect pests of agricultural and forest importance.

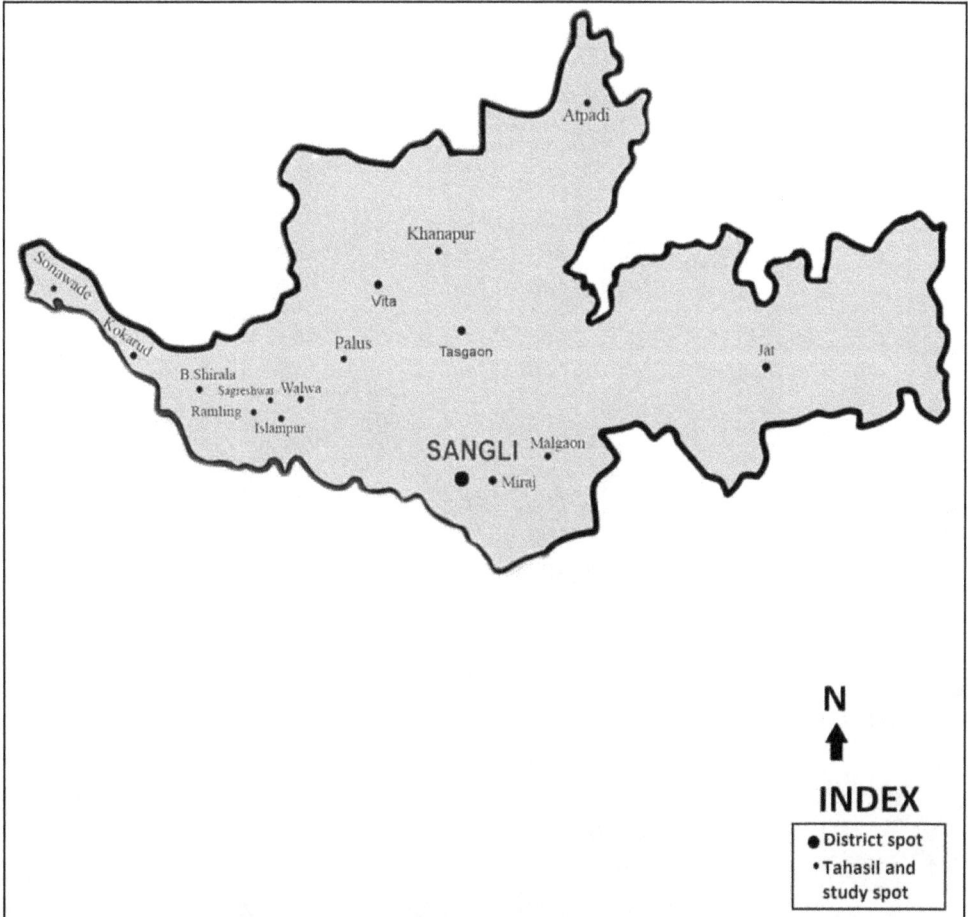

Figure 2: Map of Sangli District Showing Study Area.

Biodiversity has different concepts such as morphological, taxonomical, physiological, biochemical, cytological and molecular. Among the biodiversity concepts chromosomal diversity plays a very important role since, chromosomes carry genetic material and extremely helpful for understanding speciation. Chromosomal number, size, shape, etc plays an important role in identification of biodiversity. Hence, in the present book, chromosomal diversity has been reported in lady bird beetles and chromosomal diversity concepts in other insects have also been discussed.

2

Review of Literature

Review of literature of chromosomal diversity indicates that several workers attempted the chromosomal diversity of insects. Some notable workers are mentioned below.

Sl.No.	Year	Author	Title
1.	1866	Mulsant, M. E.	Monographic des Coccinellides Ire partie Coccinelliens pp. 1-294, Paris.
2.	1873.	Lewis, G.	Notes on Japanese Coccinellidae. *Ent. Mon. Maq.*, 10(2): 54-56.
3.	1874.	Crotch, G. R.	The revision of Coleopterous family Coccinellidae. PP 311., London (I, II).
4.	1879.	Lewis, G.	On certain new species of Coleoptera from Japan. *Ann. Maq. Nat. Hist. Lond.* (5)4: 459-467.
5.	1896.	Lewis, G.	On the Coccinellidae of Japan. *Ann. Maq. Nat. Hist. Lond.* (6) 17: 22-41.

Sl.No.	Year	Author	Title
6.	1899.	Casey, T. L.	A revision of the American Coccinellidae. *J.N.Y. Ent. Soc. 7*: 71-169.
7.	1907.	Sicard, A.	Coleopterous Coccinellides du Japan. *Bull. Mus. Hist. Nat. Paris.*, 1907: 211.
8.	1908.	Casey, T. L.	Notes on the Coccinellidae. *Can. Ent.* 40: 3493-421.
9.	1909.	Sicard, A.,	Revision des Coccinellides de la faune. *Ann. Soc. Ent. Fr.*, 78: 68-165.
10.	1915	Wheeler W. M.	On the presence and absence of cocoons among ants, the nest-spinning habits of the larvae and the significance of the black cocoons among certain Australian species. *Ann. ent. Soc. Am.* 8: 323-342.
11.	1920.	Gage, J. H.	Larvae of Coccinellidae. Illinois. *Biol. Mon.* 6(4): 232-294.
12.	1923.	Subramaniam, T. V.	Some coccinellids of South India. pp. 108-118. *Rep. Proc. Fifth Ent. Meeting, Pusa.*
13.	1924.	Aiyar, T.V.R.,	An undiscribed Coccinellid beetle of economic importance. *J. Bombay Nat. Hist. Soc.* 30: 491-493. (W.L. 25676).
14.	1925.	Arrow, G. J.	The fauna of British India, including Ceylon and Burma, Coleoptera-Clavicornia, Erotylidae, Languridae and Endomychidae, pp. 416 I Col. Pl. London.
15.	1927.	Dutt, G. R.	Aphids and lady bird beetles. *Agric. India*, 22(4). 291-292.
16.	1928	Heitz, E.	Das heterochromatin der Moose. *Jahrb. Wiss. Bot.* 69: 762-818.
17.	1931	Korschefsky, R.	Coccinellidae 1. Coleopterum Catalogues, 23(118) Den Haag. W. Junk. 224pp.
18.	1931.	Korschefsky, R.	Coccinellidae I. PP 224. Coleopterorum Catalogues Pars 118. Berlin

Sl.No.	Year	Author	Title
19.	1932.	Korschefsky, R.	Coleopteran Catalogues Pars 118. Coccinellisae I. pp. 224, Berlin.
20.	1934	Tulloch G. S.	Vestigial wings in *Diacamma. Ann. Ent. Soc. Am.* 27: 273-277.
21.	1935.	Balduf, W. V.	The bionomics of entomophagous Coleoptera (13. Coccinellidae – lady beetles). PP 220. John S. Swift Co. Inc., Chicago, New York.
22.	1936.	Anderson, W. H.	A comparative study of the labium of Coleopterous larvae. *Smithson. Misc. Coll.,* 95(13): 1-29.
23.	1941.	Binaghi, G.	Larvae Pupe di Chilocorini. *Memorie Soc. Ent. Ital.,* 20: 19-36.
24.	1943.	Timberlake, P. H.	The Coccinellidae or lady beetles of the Koebele Collection part II, Hawaii. *Plant Rec.,* 47: 1-67.
25.	1947.	Kapur, A.P.	A revision of the tribe Aspidimerini Weise (Coleoptera: Coccinellidae).
26.	1948a.	Kapur, A.P.	On the old World species of the genus *Stethorus* Weise (Coleoptera: Coccinellidae). *Bull. Ent. Res.* 39: 297-320.
27.	1948b.	Kapur, A.P.	On the Indian species of *Rodolia* Mulsant (Coleoptera: Coccinellidae). *Bull. Ent. Res.* 39: 531-538.
28.	1950.	Kapur, A.P.	The biology and external morphology of the larvae of Epilachinae. *Bull. Ent. Res.* 41: 161-208.
29.	1950.	Smith S. G.	The cyto-taxonomy of Coleoptera. *Can. Ent.* 82: 58-68.
30.	1953.	Mayne, W.W.	*Cryptolaemus montrouzieri* Mulsant in South India. Nature, 172: 185.

Sl.No.	Year	Author	Title
31.	1953.	Pattarudraiah, M. and Channabasavanna, G. P.	Beneficial Coccinellids of Mysore – I. *Indian J. Ent.* 15: 87-96. (W.L. 22997).
32.	1954.	Dyanechka, N.P.	Coccinellids of the Ukrainian SSR. pp. 156. Keiv (in Russia).
33.	1954.	Kapur, A.P.	Systematic and biological note on the San Jose scale in Kashmir with description of a new species (Coleoptera: Coccinellidae). *Rec. Indian Mus.*, 52: 257-274.
34.	1954	Ray-Chaudhari S.P. and Pyne C. K.	Time intensity factor in the production of dicentric bridges with Gamma rays of radiation during meiosis in the grass hopper, *Gesonula punctifrons*, Science, N. Y. 119: 685-686.
35.	1955.	Banks, C. J.	An ecological study of Coccinellidae associated with *Aphis fabae* Scop. On *Vicia faba. Bull. Ent. Res.* 46: 561-589. (W.L. 10184).
36.	1955.	Kapur, A.P.	Coccinellidae of Nepal. *Rec. Indian Mus.*, 53: 309-338.
37.	1955.	Pattarudraiah, M. and Channabasavanna, G. P.	Beneficial Coccinellids of Mysore – II. *Indian J. Ent.* 17: 1-5. (W.L. 22997).
38.	1956.	Pattarudraiah, M. and Channabasavanna, G. P.	Some beneficial Coccinellids of Mysore – III. *J. Bombay Nat. Hist. Soc.*, 54: 156-159.

Sl.No.	Year	Author	Title
39.	1956.	Watson, W. Y.	A study of the phylogeny of the genera of the tribe Coccinellini (Coleoptera). *Centr. Roy. Ont. Mus. Toronto (Zool.)*, 42: 1-52.
40.	1957	Virkki, N.	Structure of the testis follicle in relation to evolution in Scarabaeidae (Coleoptera). *Can. J. Zool.*, 35: 265-277.
41.	1958.	Benkevich, V. I.	Biology of *Coccinella septempunctata* Uchem. *Zap. Orekh. Zuev Pedag. Inst.*, 11: 127-133.
42.	1960.	Crowson, R. A.	The phylogeny of Coleoptera. *A. Rev. Ent.* 5(1), 111-134 (W.L. 3436).
43.	1960.	John, B. and Lewis, K. R.	Nucleolar controlled segregation of the sex chromosome in beetles. *Heredity*, 15 431-439.
44.	1961a.	Miyatake, M.	The East-Asian Coccinellid-beetles preserved in the California Academy of Sciences, tribe Hyperaspini. *Mem. Ehime Univ.* (6) 6: 67-86.
45.	1961b.	Miyatake, M.	The East-Asian Coccinellid-beetles preserved in the California Academy of Sciences, tribe Serangiini- *Mem. Ehime Univ.* (6) 6: 135-146.
46.	1961c.	Miyatake, M.	The East-Asian Coccinellid-beetles preserved in the California Academy of Sciences, tribe Hyperaspini. *Mem. Ehime Univ.* (6) 6: 147-155.
47.	1962.	Brown, W. J.	A revision of the forms of Coccinella L. Occurring in America north of Mexico (Coleoptera : Coccinellidae). *Can. Ent.* 94: 785-808.
48.	1962.	Brown, W. J., and R.de Ruette	An annotated list of the Hippodamiini of Northern America with a key to genera (Coleoptera : Coccinellidae). *Can. Ent.* 94: 643-652 (W.L. 13141).

Sl.No.	Year	Author	Title
49.	1962.	Ghani, M.A.	A note on the identity of some species of genus *Ballia* (Coleoptera : Coccinellidae). *Proc. R. Ent. Soc. London (B)*, 31: 7-8.
50.	1963.	Arnett, R. H.	The beetles of the United States. PP. 1112. Washington.
51.	1963.	Kapur, A.P.	The Coccinellidae of the third Mount Everest Expedition, 1924 (Coleoptera). *Bull. Brit. Mus. (Nat. Hist.), Ent.* 14: 1-48.
52.	1963.	Mayr E.	Animal species and evolution. Cambridge, mass: Harvard Univ. Pr.
53.	1963.	Simmonds, F. J.	Genetics and biological control problems. *Entomophaga*, 17: 251-264.
54.	1964.	Debach, Paul	Biological control of insect pests and weeds. pp. 1-843. *Chapman and Hall Ltd. London.*
55.	1965.	Chapin, E.A.	Coleoptera: Coccinellidae. *Ins. Micronesia.*, 16: 189-254.
56.	1965.	Chapin, E.A.	The genera of the Chilocorini (Coleoptera: Coccinellidae). *Bull. Mus. Comp. Zool.* 133: 227-271.
57.	1966.	Hagen, K.S.	Laboratory studies on the reproduction of *Adalia bipunctata* (Coleoptera : Coccinellidae). *Entomologia Exp. Appl.* 9: 200-204.
58.	1967.	Kapur, A.P.	The Coccinellidae (Coleoptera) of the Andamans. *Proc. Nat. Inst. Sci. India.* 32(B), 148-189.
59.	1967.	Miyatake, M.	Notes on some Coccinellidae from Nepal and Darjeeling District of India (Coleoptera). *Trans. Shikoku Ent. Soc.* 9(3): 69-78.
60.	1968	Britten, R. J. and Kohne, D.E.	Repeated sequences in DNA. *Science*, 161: 529-540.

Sl.No.	Year	Author	Title
61.	1985	Miyatake, M.	*Insecta Matsumarana*, 30: 1-33.
62.	1968a.	Sasaji, H.	Phylogeny of the family Coccinellidae (Coleoptera). *Etizenia*. 35: 1-37.
63.	1968b.	Sasaji, H.	Description of the Coccinellid larvae of Japan and the Ryukyus (Coleoptera). *Mem. Fac. Edn. Fukui Univ.* 2, (18): 93-136.
64.	1969	Clayton, F. E.	Variations in metaphase chromosome of Hawaiian Drosophilidae. Univ. Texas Pub. 6918: 95-110.
65.	1969.	Kapur, A.P.	On some Coccinellidae of the tribe Telsimini with description of new species from India. *Bull. Syst. Zool. Calcutta*, 1(2): 45-56.
66.	1969	Lahiri, M. and Manna, G. K.	Chromosome complement and meiosis in nine species of Coleoptera. *Proc. 56th Ind. Sci. Cong. Pt.*, 3: 448-449.
67.	1969	Wilson, F. D., Wheeler, M. R., Harget, M., and Kambysellis, M.	Cytogenetic relations in the *Drosophila nasuta* subgroup of the *immigrans* group of species. *Univ. Texas Publ.* 6918: 207-253.
68.	1970.	Benham, B.R., Muggelton, J.	Studies on the ecology of *Coccinella undecimpunctata* Linn. (Coleoptera: Coccinellidae). *Entomologist*, 153-170.
69.	1970	Duff, M.	The chromosomes of four New Zealand insects. *New Zealand J. Sci.*, 13: 177-183.

Sl.No.	Year	Author	Title
70.	1970	Kacker, R. K.	Studies on chromosomes of Indian Coleoptera. IV; In nine species of family Scarabaeidae. *Nucleus*, 13: 126-131.
71.	1970.	Miyatake, M.	The East-Asian Coccinellid-beetles preserved in the California Academy of Sciences, tribe Chilocorini. *Mem. Ehime Univ.* 14(2) : 303-340.
72.	1970	Ward, B. L. and Heed, W.B.	Chromosome phylogeny of *Drosophila pachea* and related species. *J. Hered.* 61: 248-258.
73.	1971.	Chapin, E.A.	The Coccinellidae of Louisiana (Insecta : Coleoptera) Doctoral Dissertation, Louisiana State University.
74.	1971.	Sasaji, H.	Fauna Japonica Coccinellidae (Coleoptera). pp. 340. *Academic Press, Japan.*
75.	1971	Wilson, E.O.	*The insect societies.* The Belknap Pess of Harvard University Press, Cambridge, USA.
76.	1972.	Kapur, A.P.	The Coccinellidae (Coleoptera) of Goa. *Rec. Zool. Survey India*, 62: 309-320.
77.	1972.	Takenouchi, Y.	A note on the cytology of *Zabrotes subfasciatus* Boh (Coleoptera: Bruchidaae). *Jap. J. Genet.* 47: 69-70.
78.	1973.	Crotch, G. R.	Revision of Coccinellidae of the United States. *Trans. Amer. Entomol. Soc.*, 4: 363-382.
79.	1973.	Pathak S., TC Hsu, FE Arrighi	Chromosomes of *Peromyscus* (Rodentia, Cricetidae). IV. The role of heterochromatin in karyotype evolution. *Cytogenet. Cell Genet.* 12: 315-326.

Sl.No.	Year	Author	Title
80.	1973	Saha, A. K.	Chromosomal studies of the Indian Coleopterans (Indian Beetles). *Cytologia,* 38: 363-373.
81.	1973.	White M. J. D.	Animal cytology and evolution. 3rd Ed. London: Cambridge Univ. Pr.
82.	1974.	Chapin, E.A.	The Coccinellidae of Louisiana (Insecta : Coleoptera). *Agri. Exper. Sta. Doyle Chambers, Bull.* 682, pp. 1-87.
83.	1975.	Weimarck A.	Heterochromatin polymorphism in rye karyotype as detected by the giemsa C-banding technique. *Hereditas.* 79: 293-300.
84.	1976.	Carson H. L., & Kaneshiro, K.Y.	*Drosophilai* of Hawaii: Systematics and ecological genetics. *Ann. Rev. Ecol. Syst.* 7: 311-346.
85.	1976.	Kitzmillar, J. B.	Genetics, cytogenetics and Evolution of mosquitoes. *Adv. Genet.* 18: 316-433.
86.	1977.	Imai, H. T., Crozier, R.H. and Taylor, R.W	Karyotype evolution in Australian ants. *Chromosoma* 59: 341-393.
87.	1977	Peacock, W.J., Lohe, A.R., Gerlach, W.L., Dunsmuir, P., Dennis, E.S. and Appels, R.	Fine structure of and evolution of DNA in heterochromatin. Cold Spring Harbor Symp. Quant. Biol. 42: 1121-1135.

Sl.No.	Year	Author	Title
88.	1978	Appels, R. and Peacock, W. J.	The arrangement and evolution of highly repeated (satellite) DNA sequences with special reference to *Drosophila*. *Int. Rev. Cytol.* 8: 69-126.
89.	1978.	Gordon, R.D.	West Indian Coccinellidae II (Coleoptera) : Some scale predators with keys to genera and species. *Coleopterists Bull.*, 32: 205-218.
90.	1978.	Smith S. G. and Virkki N.	Animal Cytogenetics 3: Insecta 5: Coleoptera. Berlin-Stuttgart: Gebruder Borntraeger X+366pp.
91.	1978.	White, M.J.D.	Modes of speciation. San Francisco: WH Freeman.
92.	1978	Yoon, J.S. and Richardson, R.H.	Evolution in Hawaii Drosophilidae. III. The microchromosome and heterochromomatin of *Drosophila*. *Evolution*, 32: 475-484
93.	1979	Hewitt, G.M.	*Orthoptera; grasshoppers and crickets*. Gerbruder Borntrager, Berlin.
94.	1979	John, B. and Miklos, G.L.G.	Functional aspects of satellite DNA and heterochromatin. *Int. Rev. Cytol.* 58: 1-114.
95.	1979	Yadav, J. S. and Pillai, R.K.	Evolution of karyotype and phylogenetic relationship in Scarabaeidae (Coleoptera). *Zool. Anz. Jena*. 202: 105-118.
96.	1981	John B.	Heterochromatin variation in natural population. In *Chromosomes Today* (ed. M.D. Benett, M. Brobow ans G. M. Hewitt), pp. 128-137. George All and Unwin, Sydney, Australia.

Sl.No.	Year	Author	Title
97.	1981.	John B.	Heterochromatin variation in natural populations. Chromosome Today 7: 128-137.
98.	1981	Peacock, W.J., Dennis, E.S., Rhoades, M.M. and Pryor, A.J.	Highly repeated DNA sequences limited to knob heterochromatin. Proc. Natl. Acad. Sci. USA. 78: 4490-4494.
99.	1982.	Baverstock PR, M. Gelder, A. Jahnke	Cytogenetic studies of the Australian rodent, *Uromys caudimaculatus*, a species showing extensive heterochromatin variation. *Chromosoma* 84: 517-533.
100.	1982	Dover, G.A.	Molecular drive: a cohesive mode of species evolution. *Nature*, 299: 111-116.
101.	1982.	Pajni, H. R. and Singh, J.	A report on the family Coccinellidae of Chandigarh and its surrounding areas (Coleoptera). *Res. Bull. Punjab Univ. Sci.* 33: 79-86.
102.	1982.	Patton JL, SW Sherwood	Genome evolution in pocket gophers (genus *Thomomys*). 1. Heterochromatin variation and speciation potential. *Chomosoma*, 85: 149-162.
103.	1982	Ranganath, H.A. and Hegade, K.	The chromosomes of two Drosophila races; *D. nasuta* and *D. albomicana*. *Chromosoma*, 85: 83-92.

Sl.No.	Year	Author	Title
104.	1983.	Gordon, R.D. and Chapin, E.A.	A revision of the New World species of *Stethorus* Weise (Coleoptera : Coccinellidae). *Trans. Am. Ent. Soc.* 109: 229-276.
105.	1983	Wakahama, K.I., Shinohara, T., Hatsumi, M., Uchida, S. and Kitagawa, O.	Metaphase chromosome configuration of the *immigrans* species group of Drosophila. *Jpn. J. Genet.* 58: 315-326.
106.	1983.	Subbarao, S.K., Vasantha, K., Adak, T. and Sharma, V.P.	*Anopheles culicifacies* complex: Evidence for a new sibling species, species C. *Ann. Ent. Soc. Am.* 76: 985-988.
107.	1984	Dover, G.A. and Flavell, R.B.	Molecular co-evolution: DNA divergence and the maintenance of function. *Cell*, 38: 622-623.
108.	1985	Cokendolphar, J.C. and Francke, O.F.	Karyotype of *Conomyrma flava* (Mc Cook) (Hymenoptera: Formicidae). *J. New York Entomol. Soc.* 92 (4): 349-351.
109.	1985.	Miyatake, M.	Coccinellidae collected by Hokkaido University expedition to Nepal, Himalaya, 1968 (Coleoptera). *Insecta Matsumarana*, 30: 1-33.
110.	1985.	Pajni, H.R. and Verma, S.	Studies on the structure of the male genetilia in some Indian Coccinellidae (Coleoptera). *Res. Bull. Punjab Univ. Sci.* 36: 195-201.

Sl.No.	Year	Author	Title
111.	1986.	Canepari, C.	On some Coccinellids of Northern India and Nepal in the Geneva Museum of Natural History (Coleoptera, Coccinellidae). *Revue Suisse Zool.* 93: 21-36.
112.	1986.	Sathe, T. V.	Biology of *Cotesia diurnii* R and N (Hym.: Braconidae) a larval parasitoid of *Exelastis atomosa* Walsingham. *Oikoassay*, 3: 31-33.
113.	1988b.	Baimai, V.	Population cytogenetics of the malaria vector *Anopheles lucosphyrus* group. *Southeast Asian J. Trop. Med. Pub. Hlth.* 19: 667-680.
114.	1988	Clayton, F. E.	The role of heterochromatin in karyotype variation among Hawaiian pictured-wing *Drosophila. Pacific Science*, 42: 28-47.
115.	1988	Myers N.	*The environmentalists* 8: 187-208.
116.	1989	Fukumoto Y., Abe T. and Taki A.	A novel form of colony organization in the "queenless" ant *Diacamma rugosum. Physiol. Ecol.,*26: 55-61.
117.	1989	Krebs C. J.	Ecological methodology. Harper and raw publishers New York.
118.	1989	Peeters, C. and Higashi, S.	Reproductive dominance controlled by multilation in the queenless ant *Diacamma Australe. Naturwissenschaften.* 76: 177-180.
119.	1990.	Holldobler, B. and Wilson, E. O.	The Ants. Harvard University Press, Cambridge, USA.
120.	1990	Katzman M.T. and G. Cale	Tropical forest preservation using economic incentives: *Bioscience*, 40: 827-832.

Sl.No.	Year	Author	Title
121.	1990	Pardue, M.L. and Henning, W.	Heterochromatin: junk or collector's item. *Chromosome*, 100: 3-7.
122.	1991	Bonaccorsi, S. and Lohe, A.	Fine mapping of satellite DNA sequences along the Y chromosome of *Drosophila melanogaster*; relationships between the satellite sequences and fertility factors. *Genetics*, 129: 177-189.
123.	1991	Peeters, C. and Billen, J.	A novel exocrine gland inside the thoracic appandages ("gemmae") of the queenless *Diacamma australe*. *Experientia*. 47: 229-231.
124.	1991	Rojers W.A.	In: Ecology and sustainable development (ed. Gopal). pp. 81-83.
125.	1991	Shaarawi, F.A. and Angus, R.B.	A chromosomal investigation of five European species of Anacaena Thomson (Coleoptera: Hydrophilidae). *Entomol. Scand.* 21: 415-426.
126.	1992	Gatti, M. and Pimpinelli, S.	Functional elements in *Drosophila melanogaster* heterochromatin. *Ann. Rev. Genet.* 26: 239-275.
127.	1992	Peeters, C., Billen, J. and Holldobler, B.	Alternative dominance mechanisms regulating monogyny in the queenless ant genus *Diacamma*. *Naturwissenschaften*. 79: 572-573.
128.	1993	Gronenberg W. and Peeters C.	Central projections of the sensory hairs on the gemma of the ant *Diacamma*: substrate for behavioral modulation? *Cell Tissue Res*.273: 401-415.
129.	1993.	Halfiter, G. and Favila, M. E.	The Scarabaeinae (Insecta: Coleoptera) an animal group for analyzing, inventorying and monitoring biodiversity in tropical rainforest and modified landscapes. *Biol. Int.* 27: 15-21.

Sl.No.	Year	Author	Title
130.	1993.	Juan C., Pons J., Petitpierre E.	Localization of tandamly repeated DNA sequences in beetle chromosomes by fluorescent *in-situ* hybridization. *Chromosome Res.* 1: 167-174.
131.	1993	King M.	Species evolution: The role of chromosome change. Cambridge University Press Cambridge, UK.
132.	1993	Peeters, C.	Monogyny and polygyny in ponerine ants with or without queens. In *Queen number and sociality in insects* (ed. L. Keller), pp. 234-261. Oxford University Press, UK
133.	1994	Irick, H.	A new function of heterochromatin. *Chromosoma*, 103: 1-3.
134.	1994	Martins, V. G.	The chromosome of five species of Scarabaeidae (Polyphaga, Coleoptera). *Naturallia*, 19: 89-96.
135.	1994.	Rozec, M.	A new chromosome preparation technique for Coleoptera (Insecta). *Chromosome Res.* 2: 76-78.
136.	1995	Le, M.H. and Duricka, D.	Islands of complex DNA are widespread in *Drosophila* centric heterochromatin. *Genetics*, 141: 283-303.
137.	1995	Zuckerkandi, E. and Henning, W.	Tracking heterochromatin. *Chromosoma*, 104: 75-83.
138.	1996	Berghella, L. and Dimitri, P.	The heterochromatic rolled gene of *Drosophila melanogaster* is extensively polytenized and transcriptionally active in the salivary gland chromocenter. *Genetics*, 144: 117-125.

Sl.No.	Year	Author	Title
139.	1996	Colamba M.S., Monteresino E., Vitturi R. and Zunino Z.	Characterization of mitotic chromosomes of the scarab beetles *Glyphoderus sterquilinus* (Westwood) and *Bubos bison* (L.) (Coleoptera, Scarabaeidae) using conventional and banding technique. *Biol. Zentralbl.* 115: 58-70.
140.	1999	Mola, L.M., Papeschi, A.G. and Taboada, C.	Cytogenetics of seven species of dragonflies. *Hereditas*, 131: 147-153.
141.	1999	Veuille M., Brusadelle A. Brazier L. and Peeters C.	Phylogenetic study of a behaviourial trait regulating reproduction in the ponerine ant *Diacamma*. in social insects at the turn of the millennium (ed. M. Schwarz and K. Hogendoorn), pp. 442. 13th Congress of the International Union for the Study of Social Insects (IUSSI), Adelaide, Australia.
142.	2000	Cherian P. T.	On the status, origin and evolution of hot spots of biodiversity. Zoo's Print. 15(4): 247-251.
143.	2000	Costa, C.	Estado de conocimieento de los Coleoptera Neotropicales. In Hacia un proyecto CYTED para el inventarion y estimatcion de la diversidad entomologica en iberoamerica; prIBES-2000. Monografias tercer milenio. (ed. F. Martin-Piera, J.J. Morrone and A. Melic), 1: 99-114.
144.	2001	Bhosale, Y. A. and Sathe, T. V.	Insect pest predators. pp. 1-167. Daya Publishing House, New Delhi.

Sl.No.	Year	Author	Title
145.	2001	Machado, V., Galian, J., Araujo, A.M. & Valente, V.L.S.	Cytogenetics of eight neotropical species of *Chouliognathus* Henzt, 1830; implications on the ancestral karyotype in Cantharidae (Coleoptera). *Hereditas*, 134: 121-124.
146.	2002	Joseph T. M., Biju A. and Shekha V.	Environmental importance of *Sacred groves* and their conservation. *Millennium Zoology*, 2 (1): 22-25.
147.	2003	Baudry E., Peeters C., Brazier L. Veuille M. and Doums C.	Shift in the behaviours regulating monogyny is associated with high genetic differentiation in the queenless ant *Diacamma Ceylonense*. *Inect Soc.* 50: 390-397.
148.	2003.	Petitpierre, E. and Gameria, I.	A cytogenetic study of the leaf beetle genus Cyrtonus (Coleoptera, Chrysomelidae). *Genetica*, 119: 193-199.
149.	2003	Gregory T. R., Nedved O., Adamovicz S. J.	C-value estimates for 31 species of Lady bird beetles (Coleoptera: Coccinellidae). *Hereditas*. 139: 121-127.
150.	2003	Moura, R.C., Sounza. M.J., Melo, N.F. and Lira-Neto, A.C.	Karyotypic characterization of representatives from Melolonthinae (Coleoptera: Scarabaeidae): Karyotypic analysis, banding and fluorescent *in situ* hybridization (FISH). *Hereditas*, 138: 200-206.

Sl.No.	Year	Author	Title
151.	2003	Vitturi, R.; Colomba, M.; Volpe, N.; Lannino, A. and Zunino, M.	Evidence for male XO sex-chromosome system in Pentdon bidins Punctatum (Coleoptera, Scarabaeoidea, Scarabaeidae) with X linked 18S – 28S rDNA clusters. *Genes. Genet. Syst.*, 78: 427-432.
152.	2003	Patil, V. J. and Sathe, T. V.	Predators and pest management. PP 1-174. Daya Publishing House New Delhi.
153.	2004.	Petitpierre, P., Kippenberg, H., Mikhailov, Y. & Bourdonne, J. C.	Karyology and Cytotaxonomy of the Genus *Chrysolina Motschulsky* (Coleoptera, Chrysomelidae). *Zool. Anz.* 242: 347-352.
154.	2004	Sathe, T. V.	Vermiculture and organic farming. pp. 1-122. Daya Publishing House New Delhi.
155.	2004	Karagyan, G., Kuznetsova, V.G. & Lachowska, D.	New cytogenetic data on *Armenian buprestids* (Coleoptera, Buprestidae) with a discussion of karyotype variation within the family. *Folia. Biol.* 52: 151-158.
156.	2004	Wilson, C.J. and Angus, R.B.	A chromosomal investigation of seven Geotrupids and two Aegialines (Coleoptera, Scarabaeoidea). *Nouv. Rev. Entomol.* (N.S.). 21: 157-170.

Sl.No.	Year	Author	Title
157.	2004	Ramaswamy, K., Peeters C., Yuvana SP., Varghese T., Pradeep H. D., Dietemann V.	Social multilation in the ponerine ant *Diacamma*: cues originate in the victims. *Insects Soc.* 51: 410-413.
158.	2005a	Bione, E. G., Camparoto, M.L. & Simoes Z.L.P.	A study of constitutive heterochromatin and nucleolus organizer regions of *Isocopris inhiata* and *Dibroctis mimus* (Coleoptera, Scarabaeidae, Scarabaeinae) using C-banding, AgNO$_3$ staining and Fish techniques. *Genet. Mol. Biol.* 28: 111-116.
159.	2005b	Bione, E.G., Maura, R.C., Carvalho R. & Souza M. J.	Karyotype C- and fluorescence banding pattern, NOR location and Fish study of five Scarabaeidae (Coleoptera) species. Genet. Mol. Biol. 28: 376-381.
160.	2006a	Baratte S., Cobb M. and Peeters C.	Reproductive conflicts and multilation in queen less *Diacamma* atns. *Anim. Behav.* 72: 305-311.
161.	2006	Macaisne, N., Dutrillaux, A.M. & Dutrillaux, B.	Meiotic behavior of a new complex X-Y autosome translocation and amplified heterochromatin in *Jumnos ruckeri* (Saunders) (Coleoptera, Scarabaeidae), Cetoniinae). *Chrom. Res.* 14: 909-918.

Sl.No.	Year	Author	Title
162.	2007	Angus, R.B., Wilson, C.J. & Mann, D.J.	A chromosomal analysis of 15 species of Gymnopleurini; Scarabaeini and Coprini (Coleoptera; Scarabaeidae). *Tijdschr. Entomol.* 15: 201-211.
163.	2007	Carbal-de-Mello, D.C., Silva, F.A. B. and Moura, R. C.	Karyotype characterization of *Eurysternus carybaeus*: The smallest diploid number among Scarabaeidae (Coleoptera, Scarabaeinae). *Micron.*, 38. 323-350.
164.	2007	Dias, C.M., Schneider, M.C., Rosa, S.P., Costa, C. and Cella, D. M.	The first cytogenetic report of fireflies (Coleoptera, Lampyridae) from Brazilian fauna. *Acta Zool.* 88: 587-589.
165.	2007	Dutrillaux, A.M., Xie, H. and Dutrillaux, B.	High Chromosomal polymorphism and heterozygosity in *Cyclocephala tridentate* from Guadalouoe;chromosome comparison with some other species of Dynastinae (Coleoptera: Scarabaeidae). *Cytogenet. Genome Res.* 119: 248-254.
166.	2007	Ferreira, A. and Mesa, A.	Cytogenetics studies in thirteen Brazilian species of Phaneropterinae (Orthroptera: Tettigonioidea; Tettigoniidae): main evolutive trends based on their karyological traits. *Neotrop. Entomol.* 36: 503-509.
167.	2007	Poggio, M. G., Bressa, M.J. & Papeschi, A.G.	Karyotype evolution in Reduviidae (Insecta: Heteroptera) with special reference to Stenopodainae and Harpactorinae. *Comp. Cytoge.* 1: 159-168.

Sl.No.	Year	Author	Title
168.	2008	Carbal-de-Mello, D.C., Oliveira, S.G., Ramos I.C. and Moura, R. C.	Karyotype differentiation patterns in species of the subfamily Scarabaeinae (Scarabaeinae, Coleoptera). *Micron*. 38. 1243-1250.
169.	2009	Amanda, P.D.A., Cabral-De-Mello, D.C., Barros E Silva, A. E. and De Moura, R.D.C.	Cytogenetic characterization of *Eurysternus caribaeus* (Coleoptera: Scarabaeidae): evidence of sex-autosome fusion and diploid number reduction prior to species dispersion. *Journal of Genetics*, 88: 177-182.
170.	2010.	Dange M.P., Ahish Rathod	Chomosome studies on four species of Scarabaeinae (Scarabaeidae: Coleoptera). *Natl. J. l. Sci.* 7 (2): 127-130.
171.	2010	Nutan Karnik, Channaveerappa, H., Ranganath, H.A. and Gadagkar, R.	Karyotype instability in the ponerine ant genus *Diacamma*. *Journal of Genetics*, 89: 173-182.

3

What is Chromosome?

Chromosomes are found in nucleus as filamentous and thick bodies. They are hereditary materials since they carry genes from generation to generation. Chromosomes (Figure 3.1) can be clearly seen during cell division. Due to their high water content, they are not seen in the active nucleus. The chromosome (Figure 3.2, 3.3) is made up of double stranded DNA in the form of a ring attached to the cell membrane at some point. It shows two cromatids joined by a centromere (Figure 3.2). The chromosomal parts include the centromere, secondary constrictions, nucleolar organizers, telomeres and satellites (Figure 3.4). The size of chromosome vary greatly in closely related genera/species or regions or in different stages of the same organism. In *Chironomous thumii thumii* there is 27 per cent more DNA than *C.t. piger*.

The DNA of cromatid is tightly bound in a helicoid manner. This DNA winds up itself as telephone cord and twisted into series of secondary coils or supercoils and accommodate in small space of chromosome.

Cromatin (Figures 3.5, 3.6, 3.8)

It shows two bands as heterochromatin and euchromatin as condensed and dispersed DNA respectively. Thus, cromatin contain adundant DNA.

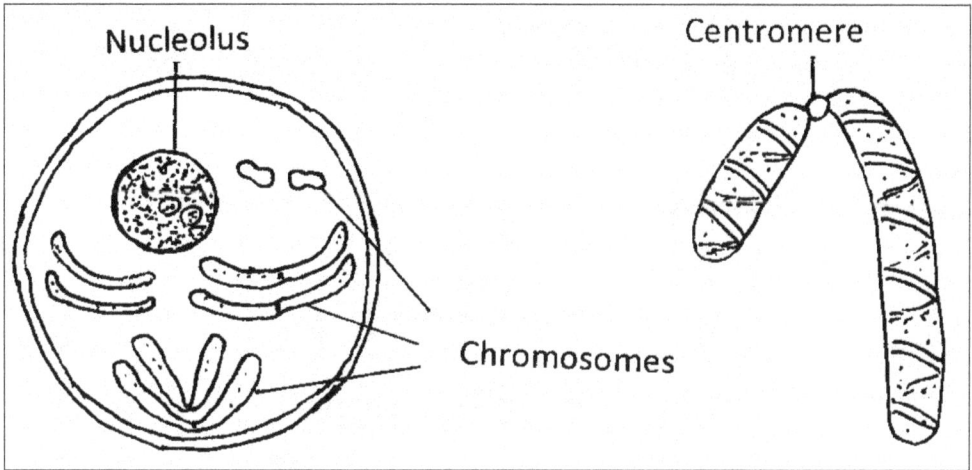

Figure 3.1: Nucleus at Cell Division with Chromosomes.

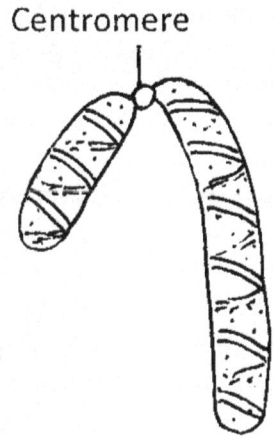

Figure 3.2: Chromosome.

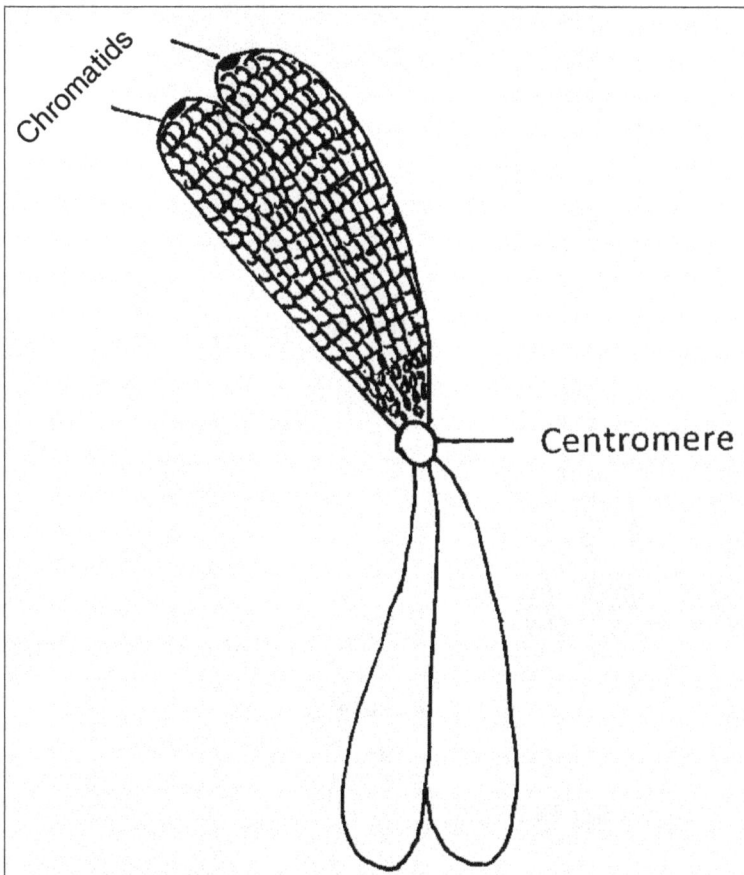

Figure 3.3: Chromosome.

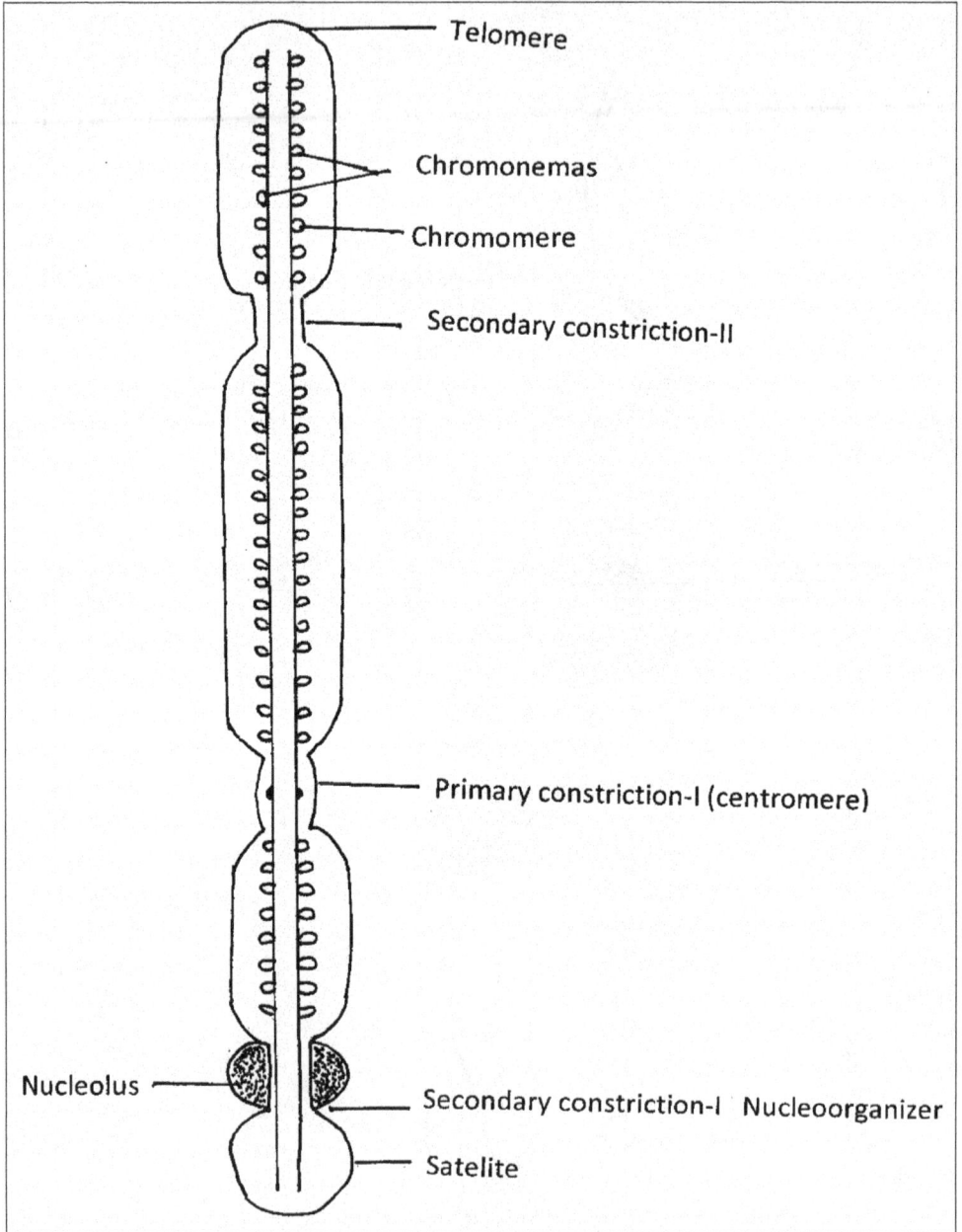

Figure 3.4: Chromosome Structure.

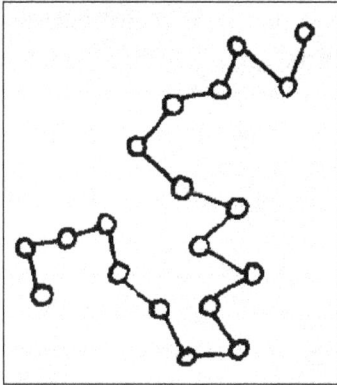

Figure 3.5: Chromatin Thread.

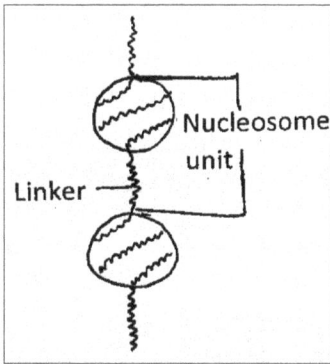

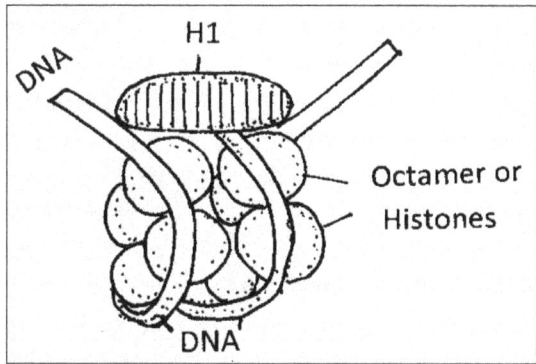

Figure 3.6: Chromatin Thread Enlarged.

Figure 3.7: Nucleosome.

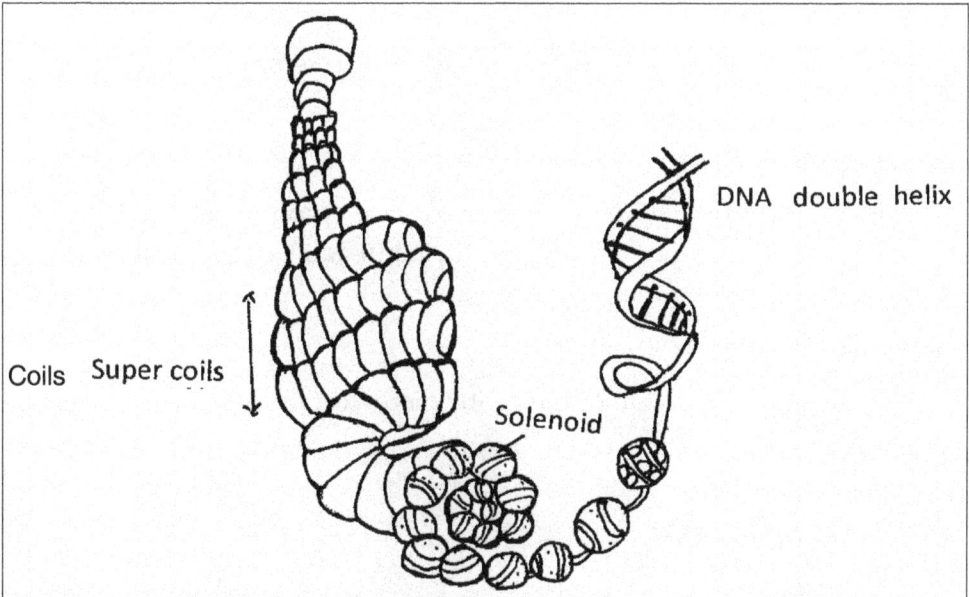

Figure 3.8: Overview Chromatin Packing.

It also contain two types of proteins, histones and non-histones. Proteins functions as organizer of chromosome and DNA acts as carrier of hereditary information. The structure of chromatin is given in the Figure - 3.3. Under electronmicroscope cromatin look like beaded material made up of repeated unit structure called nucleusomes (Figure 3.7). Each nucleosome consists of DNA and histone core.

According to Turner (1998) the classic picture of paired sister chromatids at mitosis represent the most highly condensed state of chromatin. The linear DNA traces a single path from one tip of the chromosome to the other.

Centromere (Figure 3.4)

It is the region where two chromatids are joined and also site of attachment of mitotic spindle via kinetochore, which pulls apart the sister chromatids at anaphase. Short DNA sequences is the characteristic of centromere.

Telomeres (Figure 3.4)

Telomeres are specialized DNA sequences which form the ends of the linear DNA molecules of chromosomes. A telomere consists of a short repeated sequences synthesized by a specific enzyme, telomerase.

Hetrocromatin

It is a part of chromatin in interphase which remains compact and is transcriptionally anactive.

Euchromatin

It is more diffuse region of the interphase chromosome and inactive regions in the 30 nm fiber (Figure 3.9) form but actively transcribed region for nucleosome replacement by proteins.

Types of Chromosomes

Depending on position of the centromere, four types of chromosomes are visualized.

1. Metacentric
2. Submetacentric
3. Acrocentric
4. Telocentric

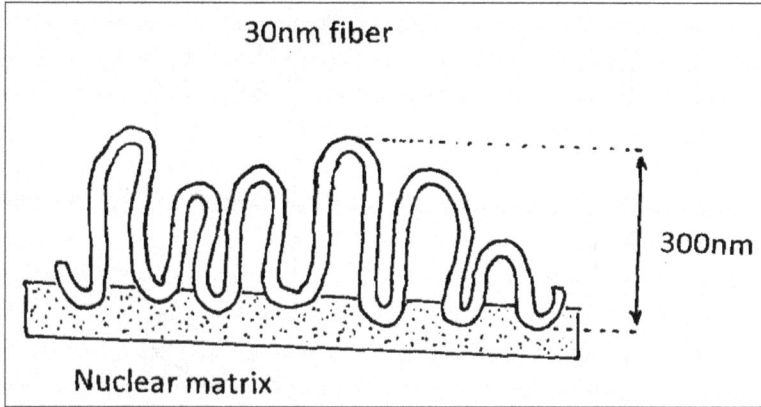

Figure 3.9: Organization of 30nm Fiber into Chromosomal Lobes.

1. Metacentric (Figure 3.10)

When centromere is at middle of the chromosome and two arms of chromosome are equal, such chromosome is called as metacentric chromosome. 'V' shaped appearance of chromosome is noticed during anaphasic movement.

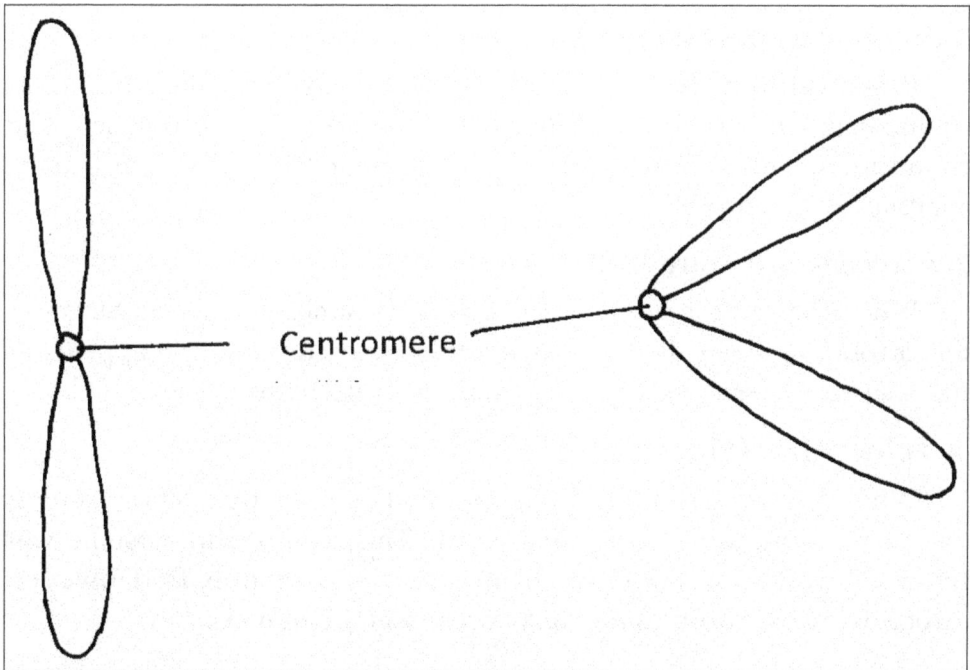

Figure 3.10: Chromosome Type: Metacentric.

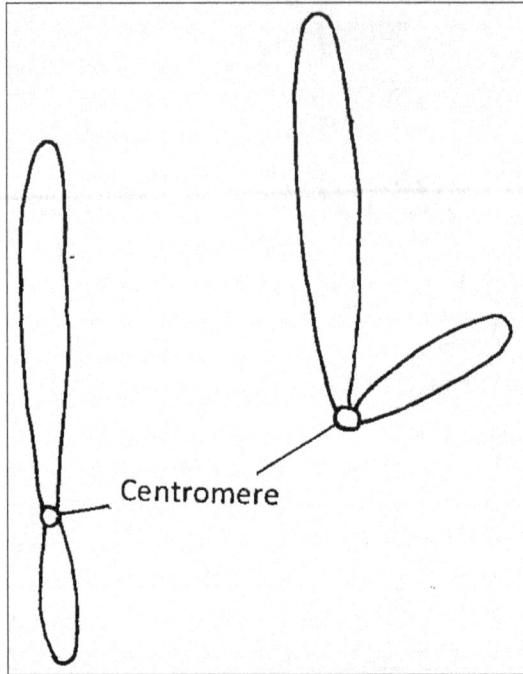

Figure 3.11: Chromosome Type: Sub-metacentric.

2. Sub-metacentric (Figure 3.11)

When centromere is at some distance away from the middle, as submedian and one arm is shorter than other, the chromosome is called as submetacentric. The chromosome appear like 'L' shape during anaphasic movement.

3. Acrocentric (Figure 3.12)

When the centromere is at the end of chromosome, as subterminal, the chromosome is called as Acrocentric. Such chromosome appears as rod shaped. Grasshopper is good example of this chromosome.

4. Telocentric (Figure 3.13)

When centromere is situated at terminal or at the tip of chromosome, the chromosome is called as Telocentric. This type of chromosome was reported by Marks (1957) in protozoans and mammals. However, acrocentric chromosomes are very rarely seen in animals.

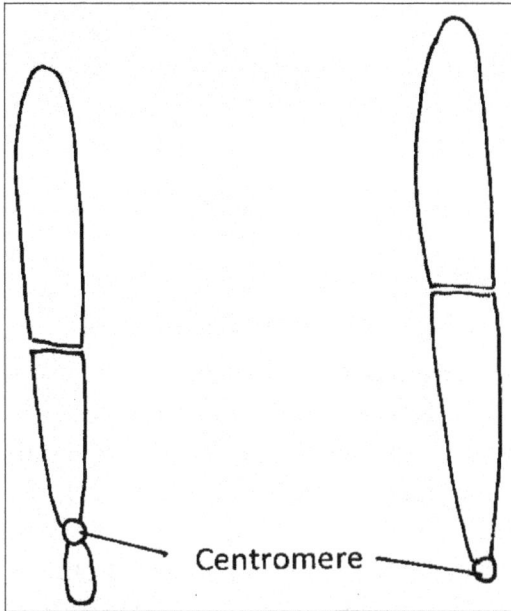

Figure 3.12: Chromosome Type: Acrocentric.

Figure 3.13: Chromosome Type: Telocentric.

5. Monocentric

This chromosome contain only one centromere.

6. Polycentric

In hemiptera and homoptera the centromere is not found in one position but lies in a diffused condition. Thus, such chromosomes are called as polycentric chromosomes.

7. Dicentric

When the chromosome is with two centromeres, the chromosome is called as dicentric chromosome. However, when two centromeres move to opposite poles during anaphase the chromosome breaks.

8. Acentric

When chromosome undergoes for a break into two resulting one centromere to one part and other part lacking. Such chromosome is called as Acentric.

9. Centromeric

Centromeric chromosome contains two chromatids and four granules within the centromere. These granules are called as Centromeric chromosomes. When two chromatids separate during anaphase, each chromatid contains four granules, these granules are called as centromeric chromosomes. During mitosis and meiosis, some time duplication of the centromere is occurred.

10. Supernumerary Chromosome (Figure 3.14)

Supernumerary chromosomes are addition to the normal autosomes and heterosomes which are genetically unnecessary. They are called accessory chromosomes or super numerary chromosomes or B chromosomes. They are not homogenous with any of the normal chromosomes. These chromosomes form synapses within themselves but not with others. Their size is smaller than normal chromosome and occurs in only some species. In more than 50 species of insects, such chromosomes are recorded, specially, acridid grasshoppers are well known for such chromosomes. They are also reported in few flat worms. However, supernumerary chromosomes are comparatively more in plants than in

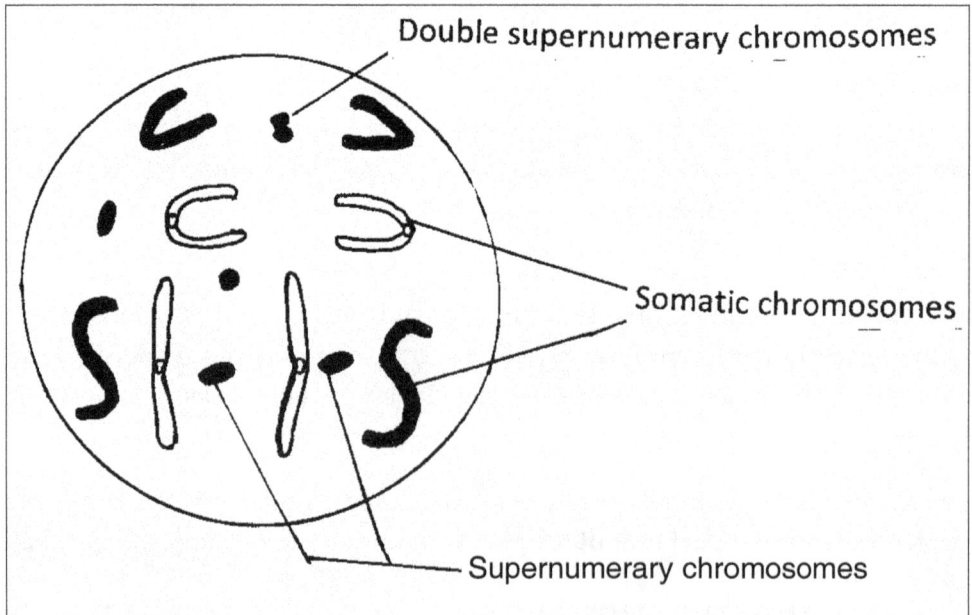

Figure 3.14: Diagrammatic Supernumerary Chromosomes.

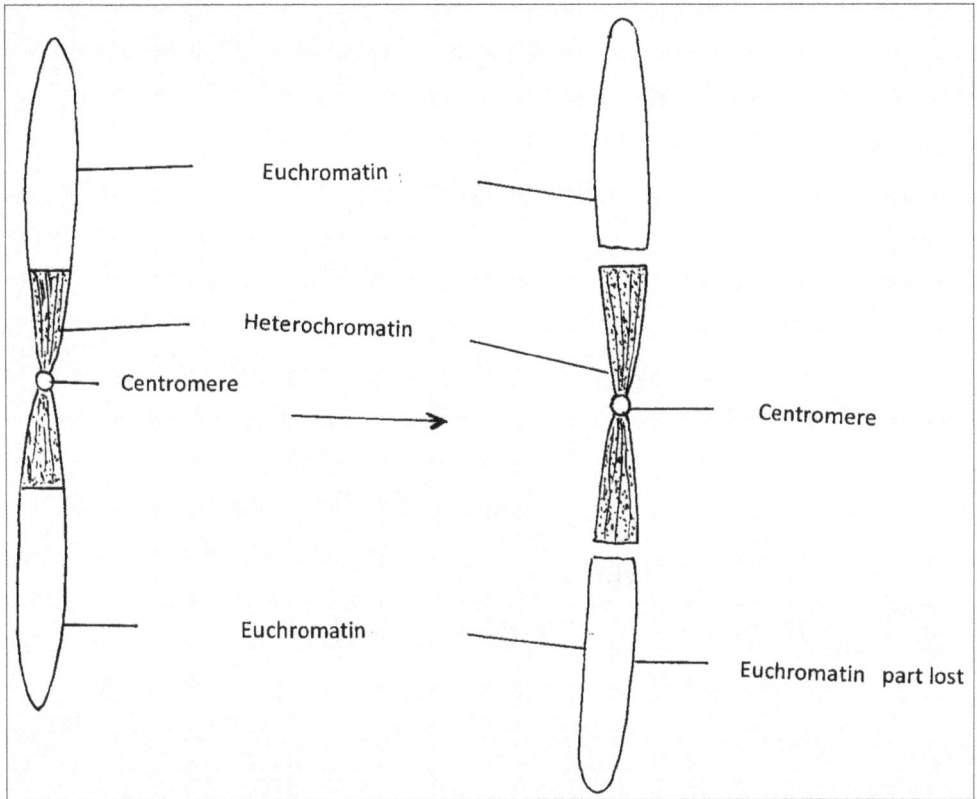

Figure 3.15: Derivation of Supernumerary Chromosome.

animals. The derivation of supermmerary chromosome is illustrated in Figure 3.15.

11. L-chromosome

These chromosomes are large but limited to the germ line and found in insects, sciarids (Diptera : Sciaridae). In the germ line cells of females of sciarid one pair of chromosome is present. In somatic cells L-chromosomes are absent. However, during the 5th and 6th cleavages these chromosomes are eliminated from nuclei but retained in germ line cells.

12. m-chromosome (Figure 3.16)

These chromosomes are extremely minute and small sized, about 0.5 micron or less than 0.5 micron. They are recorded in many Hemipterous insects specially in Coreid plant bugs (Heteroptera : Coreidae). They found

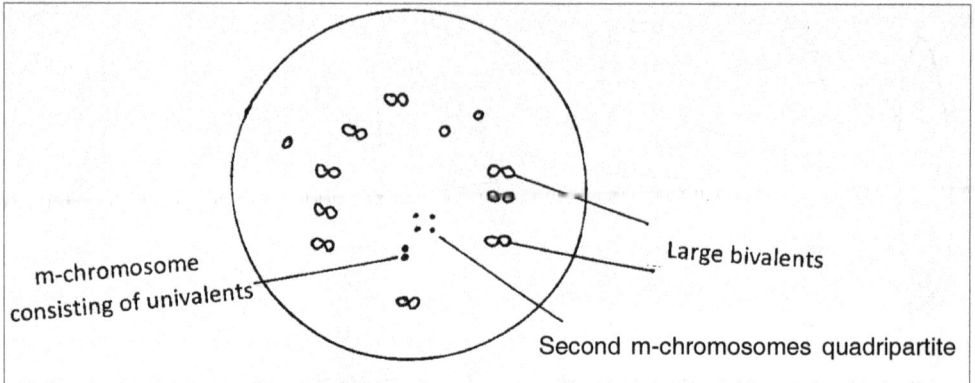

Figure 3.16: m-chromosomes.

mainly during meiosis and rarely during mitosis. Normally 1 to 2 chromosomes are seen but may reach to 5 in certain cases. m-chromosomes are also called as minute chromosomes.

13. S and E Chromosomes (Figure 3.17 and 3.18)

These chromosomes are found in some insects such as cecidomyid gall insects (Diptera : Cecidomyidae) and chironomids (Diptera : Chironomidae). In a gall insect *Maistor* sp. 48 chromosomes are found in both males and females in germ line cells. Out of which 12 are S-chromosomes and 36 are E-chromosomes. However, in somatic cells only 12 S-chromosomes have been reported in females and 4 in males. The chromosomes present in both germ and somatic cells are called as

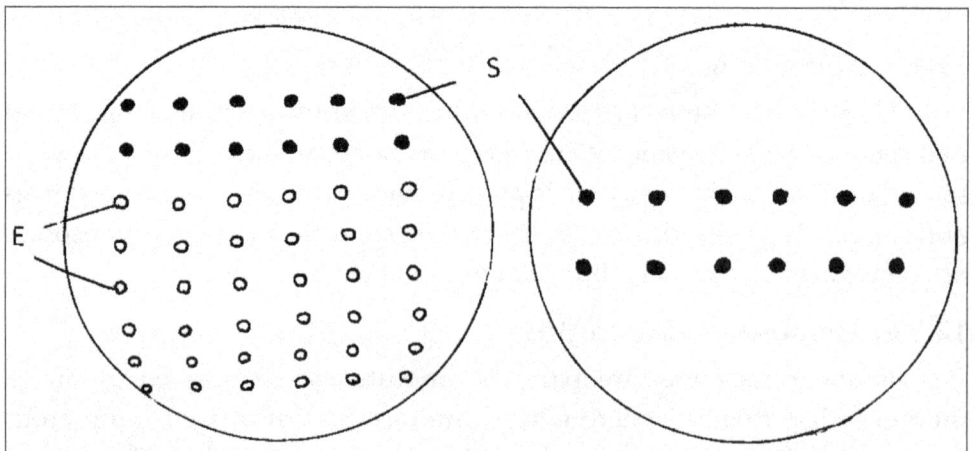

Figure 3.17: Gall Insect (Female) S and E Chromosomes S=12, E=36=48.

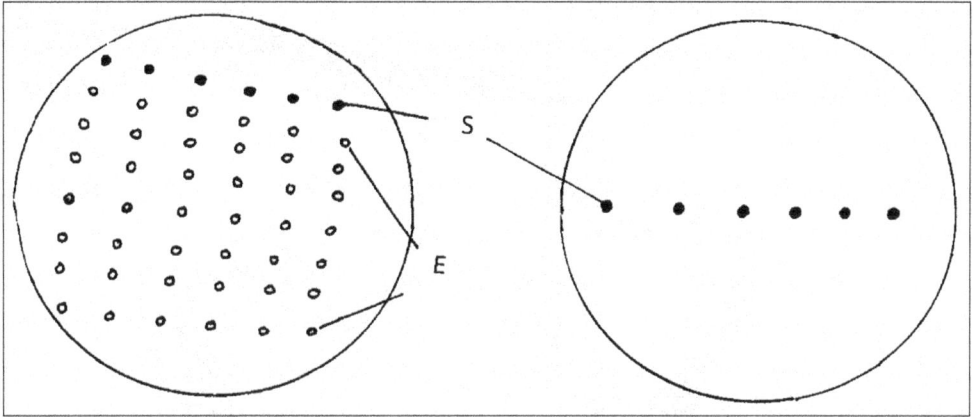

Figure 3.18: Gall Insect (Male) S and E Chromosomes E=Eliminated, S=Somatic.

S-chromosomes and those eliminated from somatic cells but present in germ cells are called as E-chromosomes.

14. Megachromosomes

These chromosomes are reported in some plants like *Nicotiana* hybrids. Generally, only one chromosome is found in a single cell - as hetrochromatic material.

15. Heterochromosomes

These chromosomes which remain condensed during interphase are called hetrochromosomes.

Sex chromosomes in insects are good examples.

16. Euchromosomes

Euchromosomes are non-condensed chromosomes, which are found during interphase.

17. Giant Chromosomes (Figure 3.19)

Giant chromosomes are polytenic consisting many strands and found in the salivary glands of Dipteran insects. There are two types of Giant chromosomes.

    i) Polytene chromosomes

    ii) Lampbrush chromosomes

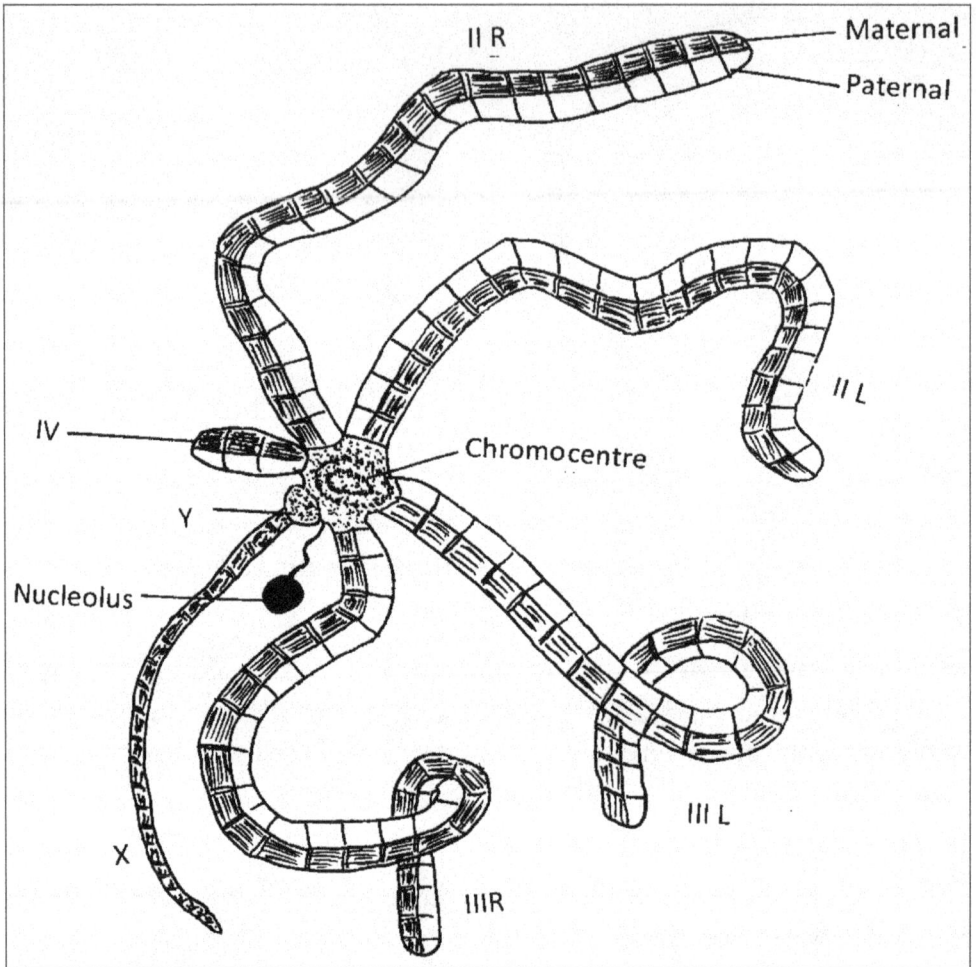

Figure 3.19: Giant Chromosomes of *Drosophila*.

(*i*) Polytene Chromosomes (Figure 3.19)

Polytene chromosomes are visible during interphase. These chromosomes were first reported in salivary gland of chironomus midge by Balbiani (1881). Hence, they are also called as salivary gland chromosomes. They contain equal number of bands with similar distribution. In *Drosophila* more than 5000 bands have been observed in four chromosomes. According to Painter and Bridges (1936) the bands of these chromosomes have a close relationship to genes and some genetic defects and mutations are related to the specific bands. The bands are the sites of genes (DNA) and it is also believed that interbands also contains many genes. DNA extends continuously through the bands and interbands.

(*ii*) Lampbrush Chromosomes

Lampbrush chromosomes are like the brushes once used for cleaning glass chimneys of lamp. These chromosomes are found in many animals such as reptiles, birds, echinodermates, mollusca, etc. and several insects. The insect chromosomes are not as large as vertebrates. They are more than 1000 m in length and 20m in width found in early prophase-I. They may contract and reduced in toward end of prophase-I. The chromosome contains chromosomal axis, chromomeres and loops. The loop continuously spins out from the chromomere at the thin end and rewinds at the thick end. On DNA axile fibre dense granules are present which has important role in RNA synthesis. The chromores are found in pairs, one for each filament.

Chromosomal Models

The Chromosomal models are grouped into

1. Multiple strand models and
2. Single strand models

The Chromosome contains DNA and protein. The arrangement of DNA and association of proteins with DNA are important aspects of consideration.

Whether DNA presents as a continuous strand from one end of Chromosome to the other or is it interrupted by linkers. The linear sequence of genes should be preserved and process of replication and segregation of DNA needs understanding.

1. Multiple Strand Models

In Multiple strand models the Chromosome should consists of several DNA - protein strands. The strands noticed are of varying thickness with fibrils ranging from 30 to 200 Å or even 500 Å. Each Chromosome consists of at least four subunits *i.e.* Chromonemata C two Chromatids, each with two halves. The model may have 4, 8, 16 and 32 strands and each strand (fibril) may contain 20 to 40 Å DNA double helix associated with protein. The mosquito prophase chromosomes contain at least 16 strands if not 32.

a) Simple Multi-Stranded Model

In this model the chromosome contains 64 double helices of DNA arranged in a parallel manner and twisted like a rope.

b) Ris-Multi-Stranded Model

In this model the 20 Å wide DNA molecule is associated with histone-protein to form 40 Å DNA - histone - nucleoprotein fibril. However, the evidences indicates that the chromosomes are not multi stranded for normal chromosome.

2. Single Strand Models

Many workers are in opinion that the chromosomes are single stranded. In fact, the chromosome is one long coiled or folded DNA molecule.

a) Centipede Model

In this model the chromosomes consists of a long protein backbone from which DNA coils branch off like the legs of centipede.

b) The Freese–Taylor Model

In this model there are two protein spines instead of one. The DNA chain is like the steps of ladder.

c) Coiled Coil Model

In this case, the chromosome is made up of only a single fibril which is tightly coiled and the coil is thrown into secondary coils. (Nebel *et. al.*).

d) Folded Fiber Model

In this model the DNA super coil is packed into a 80-100 Å Type - A fibre and the fibre is further coiled to form 200-350 Å Type - B fiber. This model was accepted by many worker and reported in many text books. Accordingly to some workers DNA itself is looped around histone beads to form nucleosomes. Hence, model of DNA - histone association proposed by DuPraw is not acceptable. Eukaryote chromosomes are made up of nucleoproteins called chromatin. The nucleoprotein consists of nucleic acid *i.e.* DNA and proteins which are mostly histones. Five classes of histones have been recorded with chromatin *viz.* H1, H2a, H2b, H3 and H4.

According to Zeuthen and Bak (1978) mitotic chromosomes are made up by the coiling of a large cylindrical fiber, the unit fibre. This fibre is formed by three levels of coiling. The first level of DNA coiling is in string of nucleosomes. The string of nuclosome is then coiled into a 300 Å diameter *Solenoid*. The solenoid is further coiled into a *Super solenoid* structure with diameter of 4000 Å and with 300 Å thick wall. The super solenoid structure is called as unit fibre.

Interchromosomal Connectives

Non-homologous chromosomes within a nucleus are interconnected by very thin DNA fibres. These connectives are structural branches of chromosomal DNA.

The nucleosome concept favours the chromosomal model as single stranded and DNA protein association.

Chromosome Proteins

Eukaryote chromosome is made up of DNA and protein. Histones are neutralizing basic - proteins in somatic cells. The chromosome consists of protein histones and several non-histone proteins (NHC-proteins) like protamines and acidic protein groups such as A1, A2, B, C and D with their own sub fractions.

Non histone proteins : The non histone proteins stimulate genetic activity while histone depress genetic activity. NHC-proteins shows variation in their structure and they are synthesized throughout the cell cycle. While histones are synthesized during the s - period. It is believed that NHC-proteins play important role in interaction of steroid hormones with target cell nuclei and also in meiotic chromosome pairing.

Histones

These small, basic, highly modified proteins are associated with DNA of eukaryotic cells. Through amino acid sequencing, DNA histones are classified into five classes:

1. H1 (H5) (Lysine very rich)
2. H2a (Lysine rich)
3. H2b (Lysine rich)
4. H3 (Arginine rich)
5. H4 (Arginine rich)

The basic amino acids such as argine and lysine are rich in histones but lacking tryptophan. Their modification occurred by acetylation, methylation and phosphorelation. Histones have very little variation in their amino acid sequences but are very highly conserved proteins.

Functions of Histones

1. Depression of genetic activity
2. Structural role in packing DNA molecules.

Most of the chromosomes in a cell are autosomes, with one or two sex chromosomes or heterosomes which carries the genes for determination of sex. In different species and thus, all the individuals of a species have same number of chromosomes or chromosomal number is generally different in closely related species have same chromosome number. Presence of whole set of chromosome is called euploidy. It may be haploids, diploids, triploids tetraploids etc. gametes normally contains only one set of chromosomes. This number is called as haploid number (n). Somatic cells usually contain two sets of chromosomes (diploid number: 2n). Triploid have three sets of chromosomes (3n) and tetraploid four sets (4n). The condition in which the chromosome sets are presents in multiples of *n* is called polyploidy. When a change in the chromosome number does not involved in entire set of chromosomes, but occured only in few of the chromosomes, the situation is called as uneuploidy. The uneuploids may be monosomics (2n - 1) trisomics (2n + 1), nullysomics (2n + 2) tertrasomics, double monosomics (2n − 1 − 1) and double trisomics (2n + 1 + 1).

The lowest haploid chromosome number recorded in eukaryotes is two mostly found in Mesotoma (flat worm) and *Ophryotrocha puerilis* (Polychaete). However, in *Ascaris megalocephala* only one chromosome was found but as compound chromosome which divides in to as many as 190 chromosomes in somatic cells. In higher plants only a few species have more than 15 chromosomes (haploid). The number of haploid chromosomes in most animals and plants lies between 6 to 25. In animals the highest chromosome number is 127 which is noticed in the hermit crab *Eupagurus schotensis*. As a polyploidy *Ophioglossum reticulatum* have show 630 chromosomal length varies from one micron (some fungi) to 30 microns (*Thrillium*). However, the salivary gland chromosomes of diptera, have be 2mm length. These chromosomes, represent a special case as they are not in a state of metaphasic contraction.

All the chromosomes of a species may have similar size (Symmetrical karyotype), or the different chromosomes may vary in size (asymmetrical karyotype). In the later case there may either the two distinct size groups (*e.g. Yucca arkansana*), or there may be a gradual series of different sizes (*e.g.* man). Chromosome size may vary greatly in closely related genera of *Trillium* shows hundred times the size of the chromosomes of *Medeola*, closely related genus. Size differences may be found in the different species of a genus *e.g.* the chromosome of *Allium porrum* are one half the size of chromosome of *Allium sativum*. Chromosome size may also vary within the species. The fly *Chironomous thumiii* has more DNA than *C.T. piger*.Size variation may also takes place in different regions or in different stages of the same organism. In the plant *Mediola e.g.* the root tip chromosomes are 50 per cent longer than the shoot tip chromosomes. In certain marine insects the chromosomes of the early blastula are smaller than those of later stage tissues.

A set of chromosomes of an individual or species is called a karyotype, in man the 23 pairs of chromosomes have been observed in somatic cells from the karyotype. Chromosomes are identified on the basis of (1) The total length of the chromosome, (2) Arm ratio (Ratio of the lengths of the long and the short arms are determined by the position of the centromere) (3) The position of the secondary constrictions and nucleolar organizers and (4) Sub division of the chromosome in to euchromatic and heterochromatic regions. Homologous pair of identified chromosomes are arranged in a series of decreasing length is called an *Idiogram*. In some species the chromosomes set may be of the same length and have their centromere located in same position. In such cases it is not possible to prepare idiograms. However, supernumerary chromosomes are not homologous with any of the normal chromosomes and do not forms synapses with the latter. They are not present in all the individuals of the species and are generally smaller than normal chromosomes. Supernumerary chromosomes are more abundent in plants than in animals. Usually each nucleus has one or two supernumerary chromosomes. There may be five supernumerary chromosomes in addition to 12 somatic chromosomes. As a result of crossing up to supernumeraries per nucleus have been obtained. Supernumerary chromosomes may be eliminated from certain tissue or organs during embryogenesis, or may be persistant in others (*Sorghum*) they may persist in the shoot, from where they transferred to the next generation.

Large chromosomes,(megachromosomes) may be 15 times longer than normal chromosomes, in hybrids, (*Nicotiana*). The giant chromosomes found in the salivary glands of dipterous insects are *polytenic*, consisting many stands. Megachromosomes are found only in a very small proportion of the total number of cells. Usually there is only one megachromosome is per one cell, although as many as 7 have been recorded in certain species.They are largely heterochromatic. They may be single centromere or may be dicentric or acentric. Megachromosomes are not transmitted through the gametes. This indicates that the ability to produce these chromosomes is inherited. Certain segments of the chromosomes, or the entire chromosomes, are more condensed than the rest of the karyotype during various stages of the cell cycle. Such difference in thickening has been called hetero-pycnosis. Heteropycnosis may be positive, or negative. Chromosomes which remain condensed during interphase are called heterochromosomes, as the sex chromosomes of insects. The non-condensed chromosomes which extend during interphase are called euchromosomes. Chromatin materials is of two types, heterochromatin and euchromatin. Chromatin material showing heteropycnosis at any stage is called heterochromatin. Regions of the Chromosomes which never show heteropycnosis consist of euchromatin.

Difference between heterochromatin and euchromatin.

1. Heterochromatin stains deeply while euchromatin stains less deeply.

2. Heterochromatin is found in the condensed regions of the chromosomes, and is associated with tight folding and coiling of the chromosome fiber. Euchromatin consists of the diffused or less tightly coiled regions. It undergoes the typical cycle of condensation during division and decondensation during interphase.

3. Heterochromatin is late replicating. It replicates at the end of the S phase of the mitotic cycle. Euchromatin replicates during the early stage of the phase.

4. Heterochomatin does not become acetylated. Euchromatin take up acetic acid (via acetyl CoA) on its histone during inter phase.

5. Heterochromatin is more labile than euchromatin and is affected by temperature, sex, age of parents, proximity to the centeromere and presence of an additional Y chromosome.

6. Heterochromatin is relatively inert metabolically. Heterochromatin segment contain relatively few genes in relation to their length but nevertheless a few genes are present.

7. The crossover frequency is less in heterochromatin than in euchromatin. Due to condensed regions of the chromosome fiber cannot come close together for frequent crossover. This mechanism may help in protecting vital genes from the effects of crossover. Chromosomatin are of two types *constitutive heterochromatin* and *facultative heterochromatin*. Constitutive heterochromatin shows heteropycnosis in all cell types. Facultative heterochromatin on the other hand is heteropycnotic only in some special cell types, or at some particular stages of the life cycle.

Constitutive heterochromatin is originally satellite DNA (S-DNA). Which is mostly inactive during protein synthesis. Constitutive DNA is highly repetitive. It consists of comparatively short identical variation in chromosome number is of two types euploidy and uneuploidy. In euploidy there is variation of entire set of chromosome. Polyploidy is comparatively rare in animals because animals have a more delicate sex balance mechanism than plants. Many species of animals, have certain tissue in their body polyploidy cells. In *Neodiprion* and *diprion* seven chromosomes were in parthenogenetically reproducing males and 14 chomosomes were present in females. *Diprion simil*, however, has14 chromosomes in the male and 28 in the female. Tetraploid tomatoes have a higher vitamin C content, and yellow tetraploid maize contains 20 per cent more vitamin than corresponding diploids. Variations in chromosome may be due to euploidy. Uneuploidy plants and animals have incomplete genomes.

Chromosomes are a definite structured and organization bodies normally found constant from one mitosis to the next. Structural modifications are known as chromosomal aberrations and chromosomal mutations. Gene mutations involve changes in only single genes on the chromosome. Chromosomal mutations on the other hand usually results in changes in blocks of genes. Chromosomal changes occur very rarely in nature. However, chemicals, X-rays and atomic radiations can play important role in chromosomal mutations.Chromosomes are interesting material from the view point of heredity and their use in species identity. Hence, in the present work chromosomal biodiversity has been studied in Lady bird beetles (Coleoptera: Coccinellidae) as Lady bird beetles are very potential bio-control agents of insect pests.

4

Collection, Preservation and Methodology

Materials

Materials and Methods play very important role in technology, any minor change in materials and methods can lead drastic change in the results both positive and negative, hence following materials and methods have been adopted.

1. Insect Collection Net (Figure 4)

Insect hand net made up of aluminum handle with 70 cm long, circular iron ring of 22 cm diameter was used for the collection of Lady bird beetles.

2. Polythene Bags (Figure 5)

Polythene bags of size 18 × 12 cm were used for collection of Lady bird beetles. During the transportation, for aeration such bags were pinned by fine points.

3. B.D. Needle Type Syringe (Figure 6)

The syringe used by diabetic patients has very fine needle which was used to inject the colchicines in to the body of the beetles.

4. Dissecting Microscope [Getner India made (SDZ-Ph)] (Figure 7)

Dissecting microscope was used for dissecting Lady bird beetle to remove the gonads and mid-gut dissection microscope.

5. Glass Cavity Blocks (Figure 8)

Glass cavity blocks of 4 X 4 X 0.5 cm (length, width and depth) were used to process the tissue through various chemical treatments for metaphasic chromosomal preparation.

6. Rubber Dropper (Figure 9)

Rubber dropper was used for picking and transferring the tissue for one grade to other grade of chromosomal technique.

7. Watch Glass (Figure 10)

The watch glass of size (6 cm diameter) used to macerate the tissue in it with the help of glass rod.

8. Glass Rod (Figure 11)

The glass rod with one flat tip was used for macerating tissue in 50 per cent acetic acid solution which caused the softening and separation of tissue.

9. Glass Slide and Cover Slips (Figure 12)

Good quality microscopic slides and cover slips were used for preparing the metaphasic chromosome by squash technique.

10. Pointed Forceps (Figure 13)

Pointed forceps of various size were used to handle the beetles at the time of dissection under the dissecting microscope.

11. Micro Scalpel (Figure 14)

Micro scalpel was prepared by cutting the razor blade and to open the abdomen of insect for the removal of testes, ovaries or mid-gut under dissecting microscope.

12. Stoppard Bottles (Figure 15)

Stoppard bottles of 100 to 500 ml capacity were used to store the chemical preparations required for the technique.

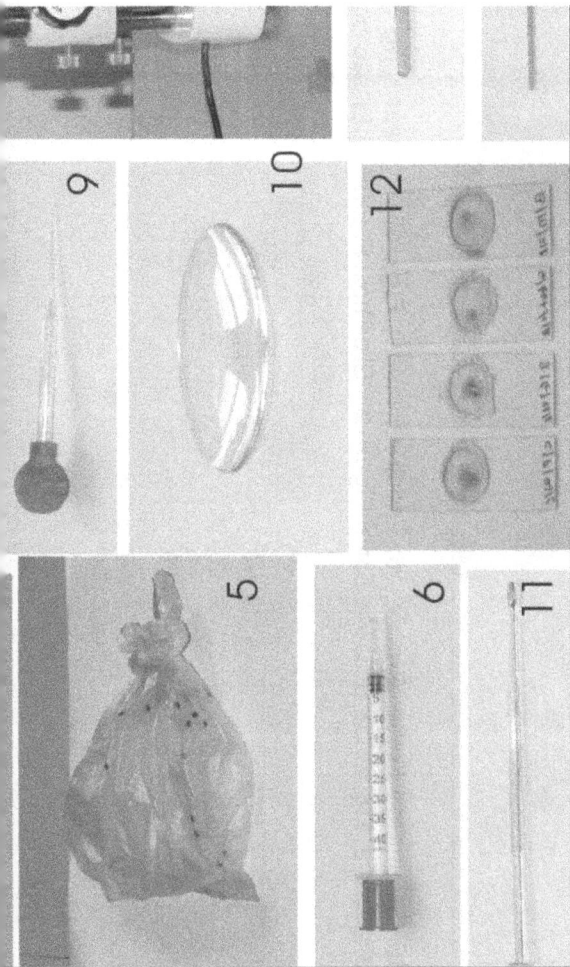

Plate 1

Figure 4: Insect collecting net; **Figure 5**: Polythene bag; **Figure 6**: B.D. needle type syringe; **Figure 8**: Glass cavity block; **Figure 9**: Rubber dropper; **Figure 10**: Watch glass; **Figure 1** cover slip; **Figure 13**: Pointed forcep; **Figure 14**: Micro scalpel.

Plate 2

Figure 15: Stoppered bottle; **Figure 16**: Drooping bottle; **Figure 17**: Measuring c
Figure 19: Camera; **Figure 20**: BOD incubator; **Figure 21**: Slide box; **Figure 22**: Dissection

13. Dropping Bottles (Figure 16)

Dropping bottles of size 50ml and 100 ml were used to store the stain and other chemicals.

14. Measuring Cylinder (Figure 17)

Measuring cylinder of 100 ml was used to prepare the chemicals required for the technique.

15. Microscope (Figure 18)

Trinoccular microscope of Magnus with objectives 5X, 10X, 40X and 100X with oil emulsion along with image analyzer camera and computer was used for observation and photography of the chromosomes.

16. Camera (Figure 19)

Camera (Sony cyber-shot 7.2 mega pixels and Canon SLR) was used for the external photography of beetles.

17. BOD Incubator (Figure 20)

BOD incubator of Lab Hosp of model and size was used to treat squashed and stained slides at 4°C.

18. Slide Boxes (Figure 21)

Slide boxes of size 28 X 22 X 3.5 cm, 21 X 19 X 3.5 cm (length, width and height) were used for keeping the slides of chromosomal preparation for avoiding dust and for safety.

19. Dissection Box (Figure 22)

Well equipped dissection box was used for the preparation of metaphasic chromosome from the tissue of beetles.

Solutions

Following solutions were used for chromosomal preparation techniques,

- ☆ 0.67 per cent NaCl: 0.67 g NaCl in 100 ml distilled water.
- ☆ 0.05 per cent Colchicine: 0.05 g Colchicine powder dissolved in 100 ml distilled water.
- ☆ 0.9 per cent Sodium Citrate Solution.: 0.05 g Sodium citrate dissolved in 100 ml distilled water.

☆ 0.56 per cent KCl Solution: 0.56 g KCl dissolved in 100 ml distilled water.

Fixatives

3:1 Absolute Methanol – Glacial acetic acid.

Stain

Aceto-orcine Solution.

Experimental Insects

Species of lady bird beetles have been used for chromosomal studies.

Methods

Survey, Collection and Identification

Survey of Coccinellid predators was made from of Sangli district during 2006-2011 by collection and spot observations of beetles by one man one hour method. A large number of specimens were collected from agricultural crops *viz.*, Jowar, *Sorghum vulgare* pers.; Maize, *Zea mays* Linn.; Groundnut, *Arachis hypogaea* Linn.; Soyabean, *Glycine soja* Linn.; Cow pea, *Vigna unguiculata* Walp.; Sun flower, *Helianthus annus*; Safflower, *Carthumus tinctorius* Linn., etc. and from non agricultural, Subabul, *Leucaena retusa*, Brew.; Rui, *Calotropis gignatea* Linn.; China rose, *Hibiscus rosasinensis* and composite garden plant, *Gaillardia bulchella* Fonger. The coccinellids were also collected from ecologically varying habitats such as agricultural fields (Figures 23–26) orchards and forests.

The lady beetles were mostly abundant from early spring to mid-summer, however, collections were made throughout the year. Specimens usually collected by sweeping or beating vegetation in cultivated and wooded areas and by hand picking.

The collected specimens were killed and pinned or preserved in 70 per cent alcohol. Specimens preserved in alcohol were dissected to study the taxonomic characters such as mouth parts, spermathica and siphon.

The genitalia of specimens was dissected with the help of dissecting needle and placed into hot solution of concentrated aqueous KOH for about 3 to 5 min. the prolonged exposure to KOH causes distortion of the genitalia. Genitalia then mounted in DPX.

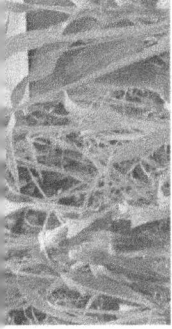

Plate 3

Figure 23–26: Author on filed spots; **Figure 26**: Lady bird beetles on Jowar crop.

Larvae of predatory coccinellidae were also collected from field and reared in the laboratory for their adult emergence. Collections were made early in the morning and evening throughout the year. After sorting of different groups and genera, each field collection was duly labeled. Antennae, labrum, mandibles, maxillae, labium and hind legs were mounted in DPX on slides for taxonomical studies, for identification of species.

Collection Area

In Maharashtra, Sangli district is leading in agriculture and industrialization which lies between 16°, 45′ and 17°, 38′ north altitudes and 73°, 42′ and 75°, 40′ East longitude. This district is bounded on East by Bijapur district of Karnataka state, on the West by Ratanagiri, on the South by Kolhapur (MS) and Solapur and Satara districts lies on the North boundaries. This district falls partly in Krishna basin and partly in Bhima basin. The whole district can also be divided in to three different parts on the basis of topography, climate and rainfall.

☆ Western hilly area of Shirala Tahsil with heavy rainfall.

☆ The basin area of Krishna and Warna and Yerala comprising of Walwa Tahsil, Eastern part of Shirala tahsil, Western parts of Miraj and Tasgaon Tahsils.

☆ An eastern drought prone area which comprises eastern part of Miraj and Tasgaon Tahsils, North Eastern part of Kolhapur and whole of Atpadi, Kavathe Mahankal and Jath tahsils.

The climate gets hotter and dried towards the east and humidity goes on increasing towards the west. The maximum temperature ranges between 31.1°C in July to 41.5°C in April, similarly the minimum temperature ranges from 10.3°C in December to 21.5°C from April to June. The rainfall and temperature have pronounced effects on climate. In Sangli district 76 major and minor irrigation projects have been launched. However, Sangli district shows less water bodies than Kolhapur.

This district have been selected on the basis of geographical and climatical parameters. Secondly, these districts have great importance in agriculture and industrialization. Hence, studying chromosomal biodiversity of coccinellid beetles from Sangli district has great economic importance from the view point of biological control of crop pests.

Morphological studies were carried out with the help of monocular microscope. Measurements were taken with the help of eyepiece micrometer and graduated mechanical stage micrometer and are in mm. The photographs of specimens were taken by using close up lenses. The identification of species was made by counsulting appropriate literature (Sasaji, 1968 and Kapur, 1963; Sathe and Bhosale, 2001; Patil and Sathe, 2003, etc.).

The terminology adopted in the present taxonomical study is the same as that of Sasaji (1968), Kapur (1970) and Chapin (1974) in the description of the species. The type material is for the time being in the collection of Department of Zoology, A.S.C. College, Palus and will be deposited to ZSI, Kolkatta. Terms used in description of species for head, antennae, mandible, maxilla, labium, hind leg, spermatheca, siphon, tegmen and male genitalia are given below. The references consulted for present work are listed in bibliography.

Biodiversity of Insects

Biodiversity of coccinellid beetles have been studied in defferent spots of study area at 15 days interval and through net swept method by man one hour collection method. The collected materials were brought to the laboratory and identified by consulting Mulsant (1846), Sathe and Bhosle (2001), Miyatake (1961), Sajaji (1968) etc.

Chromosomal Biodiversity

For chromosomal studies mitotic and meiotic metaphases are required. To study the mitotic and meiotic metaphasic chromosomes, the old squash chromosomal technique and squash chromosomal technique (Ray Chaudhari and Pyne, 1954) was used. Hypotonic colchicine treatment was given to adult insects to increase number of meiotic and mitotic metaphasic cells. Coccinellid beetles have been collected from different fields as mentioned in methods section, the collected coccinellid beetles were kept time being for further studies in small plastic containers along with their natural food *i.e.* Aphids, Delphacids, mealy bugs etc. later lady bird beetles were used for chromosomal studies. The brief steps of which are given below,

☆ Injection of 0.01 to 0.03 ml of 0.05 per cent colchicines to the beetles and keeping them for 6 to 7 hours at room temperature.

☆ Dissection of testes, ovaries and mid-gut separately in 0.067 per cent NaCl.

☆ Transfer of the tissue to 0.56 per cent KCl for 15 to 30 min.

☆ Treatment to the tissue with 0.09 per cent Sodium citrate for 60 to 90 min.

☆ Transfer of the tissue to freshly prepared fixative, allow tissue fixing to 15 to 30 min. material may be stored in this fixative for 2 to 3 months at 4°C.

☆ Transfer of the fixed material to 15 per cent acetic acid to soften the tissue for 5 to 10 min.

☆ Softening of tissue with 50 per cent Acetic acid in watch glass and macerating with glass rod.

☆ Macerating tissue on the slide with the help of pipette, remove acetic acid with the help of blotting paper.

☆ Stain the tissue with Aceto-orcine for not more than 5 min. put the cover slip; press it gently with thumb finger to spread the tissue.

☆ Bloting of the slide to remove the excess stain, seal the cover slip with the help of nail paint, protect the slide from dust and store at 4°C for overnight.

At least 5 plates were checked to count chromosome number for the calculation of the mean and mod of chromosome. The slides for good chromosomal definition were photographed at higher magnification for counting the chromosome number and to measure length of chromosome and arm of chromosome.

Chromosomes with good definition were measured. The simple and common method is to draw the chromosome at the higher magnification with the help of camera lucida for recording outline of the chromosome and their centromeric position. The length of each chromosome or arm of chromosome measured using graphics table linked to micro-computer. Arm length of chromosome then converted in to absolute units by reference to the scale bar on the drawing. Length of chromosome or arm-length of chromosome was measured by photoghraphing the chromosomes to get clear image which subsequently be enlarged. Chromosome index was prepared by following formula:

$$\text{Centromere index} = \frac{\text{Length of short arm}}{\text{Chromosomal relative length}} \times 100$$

5

Morphological Diversity of Coccinellids

Introduction

Taxonomy becoming an attractive profession due to interlinking of cytological and molecular aspects, as molecular technology is glorious field for research. Taxonomists are needed in universities, research Institutes, Museums, Central and Govt. agencies, Industries, Zoos, etc. They can play a wider role in the fields of public health, environmental problems, national defense, wild life management and pest management in identifying the species. They are also involved in designing and implementing biotechnical programmes for disease management most effectively. The success or failure of biological programme is dependent on correct identification of organisms. Therefore, increasing efforts have been directed towards the chromosomal aspects for studying biodiversity.

In the biological insect pest control programmes coccinellids are widely used. The Coccinellid beetles are familiar to people because of their prettiness and scientific importance. More than 3,50,000 species of Coleopteran species have been estimated from different parts of the World and more than 17431 species have been described from India (Sathe and Bhosale, 2001). Among the various families of Coleoptera, Coccinellidae has great importance since most of the species are bio-control agents of insect pests and hence widely used in biological pest control (Patil and Sathe, 2003). Coccinellids feed on larval and adult stages of aphids, mealy bugs, scale insects, psyllids and phytophagous mites, etc which are injurious to various agricultural and forest crops. From the viewpoint of evolutionary biology, biogeography, population ecology, genetics, cytology and biotechnology Coccinellids have great importance. According to Sasaji (1971) about 490 genera and 4200 species of Coccinellidae have been described from the World.

Linnaeus (1758) very first described 36 species of the genus *Coccinella*. Redtenbacher (1843) divided the family Coccinellidae into aphidophagous and phytophagous groups. Mulsant (1850) established his system of identification for all the Coccinellids of the World. He divided Coccinellidae into "Gymnosomides" and "Trichosomides" based on presence or absence of dorsal pubescens. In 1874, Crotch thoroughly revised the family Coccinellidae. Later, Chapius (1876) revised the family Coccinellidae into two sub divisions namely "Aphidophagous" and "Phytophagous". Further, he divided the aphidophagous into 13 groups. In 1899, Casey revised the American Coleoptera, he separated the family Coccinellidae into two major parts, Aphidophagous and phytophagous, wherein he arranged 16 tribes but not any sub-families. His arrangement of the tribes was as follows: Hippodamiini, Coccinellini, Psylloborini, Scymnillini, Hyperaspini, Cranophorini, Epilachnini, Pentilini, Chilocorini, Platynaspini, Telsimiini, Pharini, Oeneini, Scymnini, Rhyzobiini and Doccidulini. He has described a very large number of species from the World including North America.

Sicard (1907, 09) made phylogeny of the family Coccinellidae. Korschefsky (1931, 32) arranged all the Coccinellidae of the world by following system:

Epilachinae   10 genera

Lithophilinae   *Lithophilus*

Coccinellinae

Coccidulini   23 genera (including Rhizobiini),

Noviini	5 genera,
Ortaliini	17 genera,
Scymnini	20 genera (including Stethorini),
Scymnillini	*Scymnillus, Zagloba,*
Aspidimerini	*Aspidimerus, Cryptoqonus, Cyrema, Hypocyrema,*
Cranophorini	8 genera,
Hyperaspini	8 genera,
Pharini	20 genera including Sticholotini, Coleopterini, Telsimiini, Serangiini
Oeneini	9 genera,
Clanini	*Jauravia*
Pentiliini	*Pseuodosmilia, Pentilia*
Exoplectrini	10 genera (including *Chnoodini*)
Azyini	*Azya, Iadoria,*
Platynaspini	*Platynaspis, Boschalia*
Chilocorini	15 genera
Synonychini	30 genera
Coccinellini	38 genera (including Hippodamini, Anisostictini),
Psylloborini	13 genera and
Discotomini	5 genera

Watson (1956) also worked on the phylogeny of Coccinellidae almost similar to the Korschefsky's one. Some Authors (Arnett, 1962, Kapur 1963, Zaslavskij 1965) followed the Koreschefsky's key and revision with slight modification.

In 1968, Sasaji provided the following system of classification for

Family Coccinellidae

Sub family – Sticholotinae – Sukhumhikonini, Serangiini, Sticholotini, Shiroznellini

Sub-family – Scymninae   - Stethorini, Scymnini, Aspidimerini, Hyperaspini, Ortaliini

Sub – family – Chilocorinae – Telsmiini, Platynaspini, Chilocorini,

Sub-family – Coccinellinae – Lithophilini, Coccidulini, Exoplectini, Noviini,

Sub-family – Coccinellinae – Coccinellini, Psylloborini and

Sub-family – Epilachninae.

Several workers (Ahmed 1973, Chapin 1984, Gorden 1989, Bhasker, 1992, etc.) supported the system of Sasaji.

From India Subramaniam (1923) gave a list of some Coccinellids with their perys from South India. Kapur (1948b) revised the old world species of the genus Stethorus Weise and added 15 new species from India. Puttarudriah and Channabasavanna (1953, 1955, 1956) recorded 53 species in 23 genera grouped in 8 tribes and 5 sub families. Usman and Puttarudriah (1955) listed 48 species of predaceous Coccinellidae from Mysore state.

Pajni and Singh (1982) recorded 30 species of Coccinellidae belonging to 18 genera and 2 sub families from Chandigarh. Pajni and Verma (1985) also described 25 species of Coccinellids from Chandigarh. Canepari (1986) reported 36 species of Coccinellidae of northern India and Nepal in Geneva Museum. Bhasker (1992) redescribed 31 Coccinellid spp. in the form of thesis (M.Sc.) from Bangalore.

Recently sathe and Bhosale 2001, Patil and Sathe (2003), Sathe (2006) etc worked on biodiversity of lady bird beetles from Maharashtra.

They described 26 new species from Maharashtra. The family Coccinellidae is characterized by: body oval and round more or less strongly convex; antennae more or less clavate; terminal segment of maxillary palpi usually enlarged and triangular (Securiform), elytra not at all truncate and never distinctly straight; hind coxae strongly transverse; abdomen with accurate femoral lines; tarsal formula usually cryptotetramerous; siphon of male genetilia always elongate and more or less curved ventrally; tegmen trilobed.

Under the family coccinellidae following 6 sub-families are included, Sticholotinae, Scymninae, Chilocorinae, Coccidulinae, Coccinellinae and Epilachninae.

The sub family Coccinellinae is characterized by, body medium to large size; dorsal surface glabrous, antennae long, 11 segmented and inserted more or less dorsally; mandible with basal tooth, bifid apically; apical segment of maxillary palpi distinctly securiform; abdomen always composed of 6 visible sterna; femora elongate and not flattened and tarsi always cryptotetramerous and contains two tribes *viz.*, Coccinellini Weise, 1885 and Psylloborni Casey, 1899.

Tribe Coccinellini Weise, 1885.

Coccinellids Leach, 1815. Brewstar, Edingh, Encyclop., 116.

Coccinelliens Mulsant, 1846, Securipalp: 28.

Coccinellini Weise, 1885, Best, Tab. Europ. Coleopt. 11, ed. 2 : 2.

The tribe Coccinellini characterized by, body large to medium sized, usually short oval or hemispherical, rarely elongated oval; dorsum always glabrous; antennae at least nearly as long as inter-ocular distance, eleven segmented and weakly but distinctly clavate; apical segment of maxillary palpus usually securiform and mandible with a bifid tip and a basal tooth.

22 genera are included under this tribe.

Tribe Psylloborini Casey 1899.

Psylloborini Casey, 1899, J.N.Y. Ent. Soc. 7: 93, 100.

The tribe Psylloborini shows following features,

1. Mandible with multidenticulated apex, rarely with a bifid apex,
2. Anterior margin of pronotum very weakly and accurately concave,
3. Galea of maxilla very broad.

The tribe Psylloborini is closely related to the tribe Coccinellinini which contain 3 genera namely, *Vibidia Mulsant,* 1846; *Illeis Mulsant,* 1850 and *Halyzia Mulsant,* 1850.

Sub-family – Chilocorinae

This sub family is featured by, Clypeus distinctly broadly and laterally expanded; antennae reduced in the length and the segments; terminal segment of maxillary palpi never strongly divergent apically and elytral base distinctly broader than pronotal base.

Sub family Chilocorinae is divided into three tribes, namely, Chilocorini, Mulsant, 1846; Platynaspini Mulsant, 1846 and Telsimiini Casey, 1899.

Tribe Chilocorini Mulsant, 1846.

Tribe Chilocorini Mulsant, 1846, Securipalp: 166.

1. Dorsum glabrous, rarely pubescent,
2. Femora normal, not strongly depressed,
3. Tarsi cryptotetramerous and
4. Abdomen composed of 5 visible segments in female and 6 in male.

This tribe contains 2 genera, *Chilocorus* Leach, 1815 and *Arwana* Leng, 1908.

Sub family – Scymninae

This sub family is characterized by following characters;

1. Body usually small to medium sized
2. Antennae usually very short
3. Terminal segment of maxillary palpi usually nearly parallel sided and
4. Elytral base not much broader than pronotal base.

Five tribes are included under this sub family namely, Hyperaspini Mulsant, 1846; Scymnini Costa, 1849; Ortalini Mulsant, 1850; Aspidimerini Weise, 1900 and Stethorini Dobzhansky, 1924.

Tribe – Scymnini Costa 1849.

Scymniens Mulsant, 1846 Securipalpes : 20

Scymniens Costa, 1849 p. 9.

The tribe Scymnini shows, Body less than 4 mm; body shape elongate oval to hemispherical; clypeus not widely and strongly expanded laterally; terminal segments of maxillary palpi parallel sided or slightly divergent apically; anterior margin of prosternum flat or slightly concave and Elytral epipleura very narrow, horizontal and without foveae.

The tribe Scymnini contains 7 genera namely, *Scymnus* Kugelann 1794; *Nephus* Mulsant, 1846; *Cryptolaemus* Mulsant, 1853;*Horniolus* Weise, 1900; *Pseudoscymnus* Chapin, 1962; *Axinoscymnus* Kamiya, 1963 and *Keiscymnus* Sasaji, 1971.

Morphological Considerations

Adult Dorsal View (Plate 4, Figure 27).

1. Head
2. Antennae
3. Eye
4. Pronotum
5. Humeral callus
6. Scutellum
7. Elytra
8. Elytral suture
9. Fore leg
10. Middle leg
11. Hind leg
ES. Elytral spots

Adult Ventral View (Plate 4, Figure 28)

1. Hypomeron
2. Prosternal carina
3. Prosternum
4. Prosternal process
5. Postcoxal process
6. Fore coxal cavity
7. Mesosternum
8. Middle coxal cavity
9. Metepisternum
10. Femoral line
11. Metepimeron
12. Hind coxal cavity
13. Lateral femoral line
14. Femoral line
15. Elytral epipleuron

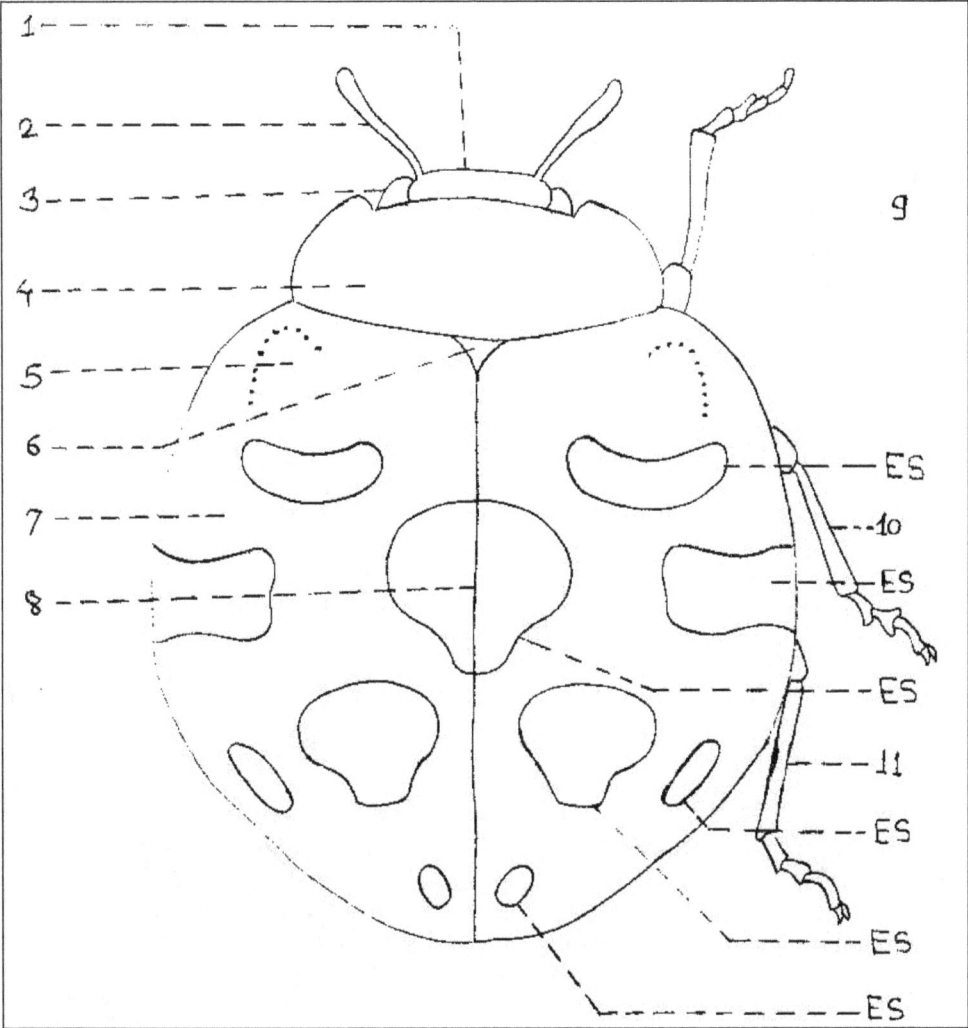

Figure 27: Dorsal view- Coccinellid.

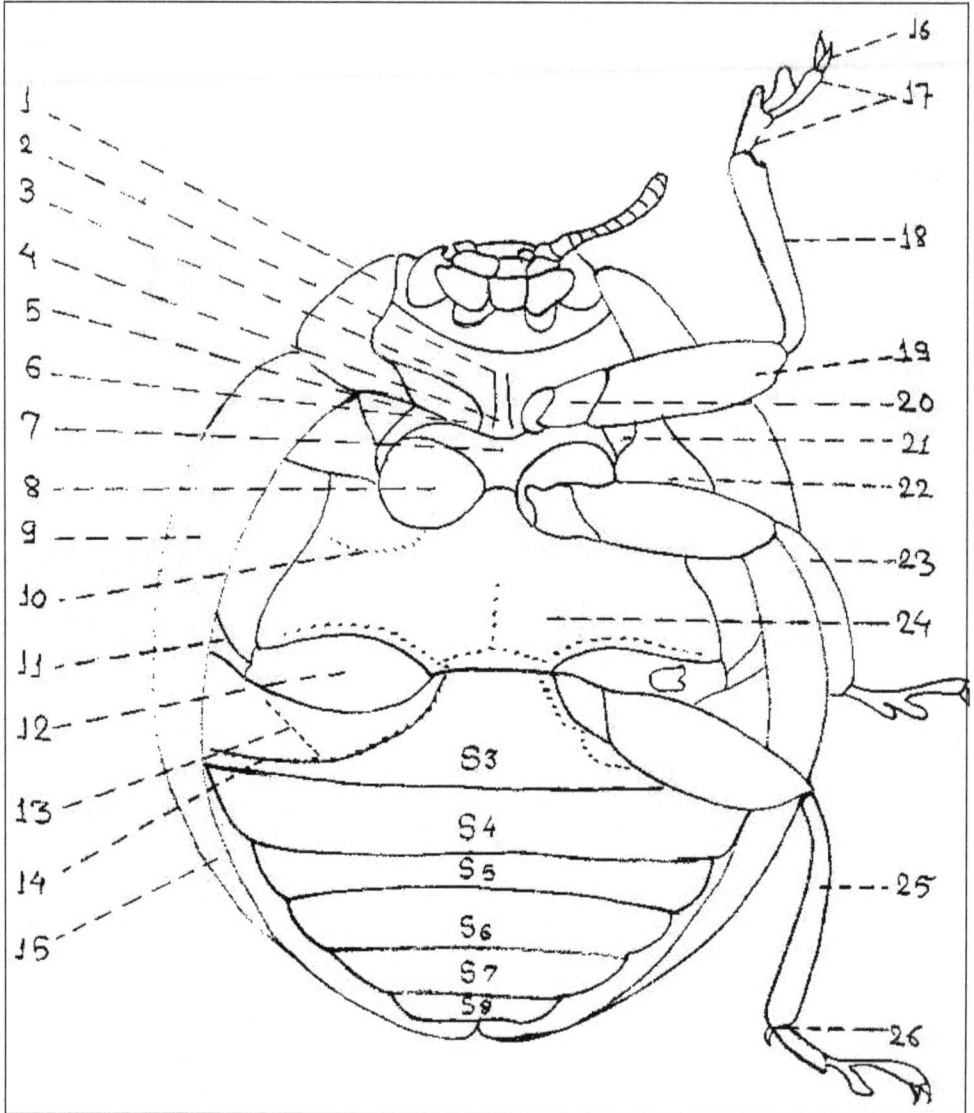

Figure 28: Ventral view- Coccinellid.

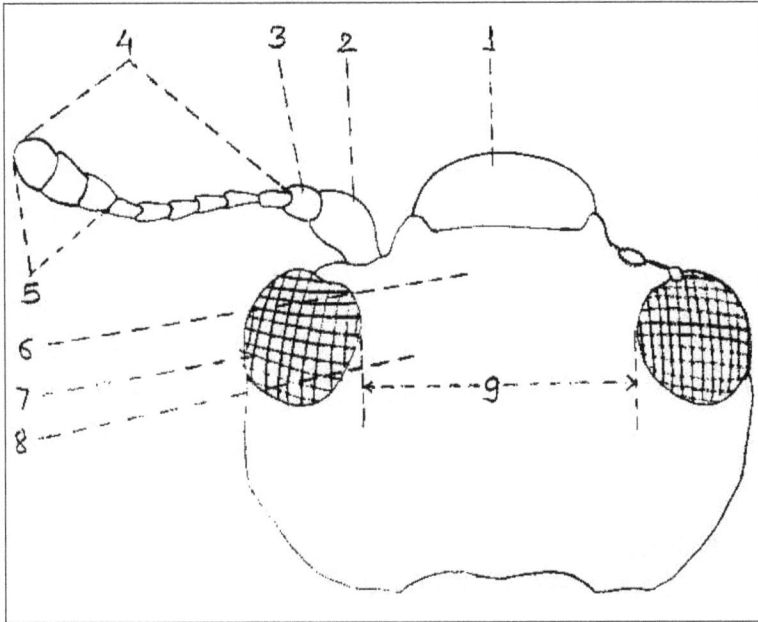

Figure 29

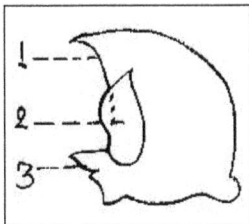

Figure 30

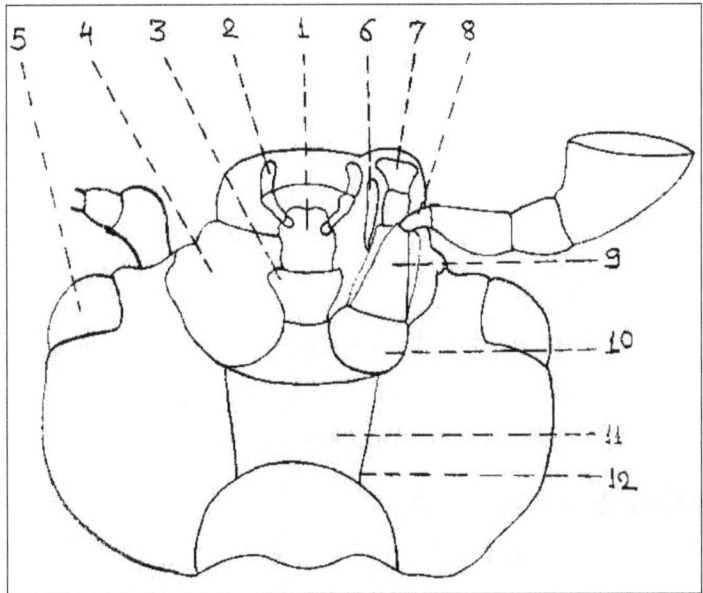

Figure 31

16. Claw

17. Tarsus

18. Tibia

19. Femur

20. Trochanter

21. Mesepisternum

22. Mesepimeron

23. Tibia of middle leg

24. Metasternum

25. Tibia of hind leg

26. Tibia spur of hind leg

S3-S8: Sternite of abdominal segments

Head Dorsal View (Figure 29)

1. Labrum

2. Scape of antenna

3. Pedicel of antenna

4. Flagellum of antenna

5. Antennal club

6. Clypeus

7. Compound eye

8. Frons

9. Interocular distance

10. Antennal socket

11. Post antennal process

Mandible (Figure 30)

1. Apical tooth

2. Membranous prostheca

3. Basal tooth

Head Ventral View (Figure 31)

1. Prementum of labium
2. Labial palp
3. Mentum
4. Tentorium
5. Compound eye
6. Lacinia of maxilla
7. Galea of maxilla
8. Maxillary palp
9. Stipe of maxilla
10. Cardo maxilla
11. Gula
12. Gular suture

Metathorax (Figure 32)

1. Metathorax
2. Postnotum of metathorax
3. Metendosternite
4. Anterior tendon of metendosternite
5. Lateral arm of metendosternite
6. Lateral lobe of metendosternite
7. Sternum
8. Coxa of hind leg
9. Metasternum
10. Metepimeron

Tegmen of Male (Figure 33)

1. Tegmen strut
2. Basal piece
3. Median lobe
4. Groove
5. Lateral lobe

Male Genitalia (Figure 34)

1. Apophysis
2. 9th tergite
3. 9th pleurite
4. 9th sternite
6. 10th tergite

Sipho (Figure 35)

1. Siphonal capsule
2. Outer lobe of siphonal capsule
3. Ejaculatory duct
4. Inner lobe of siphonal capsule
5. Trachea
6. Siphonal tube

Ovipositor and Spermatheca (Figure 36)

1. Receptaculum seminis
2. Nodulus
3. Ramus
4. Cornu
5. Accessory gland
6. Sperm duct
7. Infundibulum
8. Bursa copulatrix
9. 9th pleurite
10. Hemisternite
11. Stylus
12. 9th tergite

The persistence or recurrence, of the perticular morphological and physiological characteristics of living organisms is attributable partly to the relative constance of their physical and chemical environment and partly to their innate capacity for self maintenance and self reproduction.

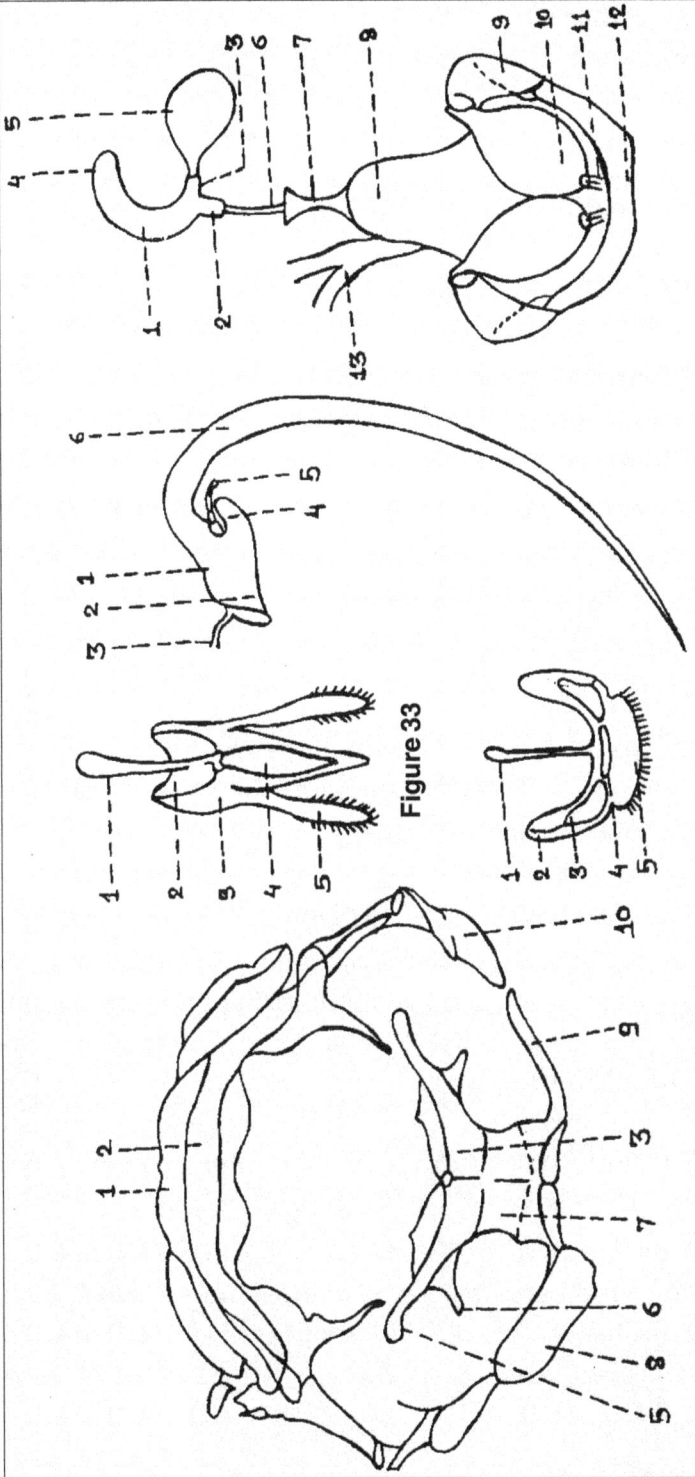

Figure 32

Figure 33

Figure 34

Figure 35

Figure 36

It is striking fact that individual living organisms do not simply maintain themselves indefinitely. Every individual cell or organism eventually dies, or else undergoes a process the result of which the production of a larger number of cells or organisms than were present initially. Given the right condition, the vast majority of these are fully capable of maintaining or recreating all of the morphological and physiological features characteristic of the particulor kind of organism. The process by which new organisms are produced can be traced in every case to the division of individual cells, involving the transmission to each daughter cell of materials necessary for growth, differentiation and farther division. Chromosome plays a crucial role in transmitting hereditary material of the organisms and related to sex determinations and species identification.

The present chapter deals with the chromosomal diversity of Coccinellidae from Sangli district of Maharashtra, India. This is the first attempt on Coccinellidae from this region. The study covers morphological and chromosomal diversity of the species, distribution and host records. Chromosomal diversity is studied in 15 species of the family Coccinellidae. Chromosomal diversity can avoid external morphological confusion and will be helpful in correct identification of the species. Hence, present study was undertaken. In past, chromosomal diversity in insects have studied by Angue (1982,1989), Crozier (1968), Gregory *et al.* (2003), Hawkins (2000, 2001), ISG (1960), ISCN (1985), John and Lewis (1960), Juan *et al.* (1993), Baimai (1980), Kanda et al.,(1983), Wibow *et al.* (1984), etc.

Chromosomal aspect on the Coccinellidae is reviewed by Smith and Virkki (1978). They gave data for about 150 species. Examination of the Zoological Record from 1978 – 2012 has revealed chromosomal data on fewer than 10 additional species, with nothing on either *Coccidula* or *Rhyzobius*. However, Gregory *et al.* (2003) gave genome-size estimates for *Rh. Litura* (F.) and *C. rufa* (Herbst), based on image analysis of Feulgen – stained nuclei. They gave no karyological information.

The chromosomes of Coccinellidae shows considerable diversity in both autosome numbers and sex chromosome systems.The perusal of literature indicates that very little attention is given on chromosomal diversity of Coccinellidae.

The present chapter deals with the morphological diversity, species, distribution and host records of coccinelllidae from Sangli districts of Maharashtra (Figure 3) India. This is the first attempt of Coccsinellidae from this region. Chromosomal diversity is studied in 15 species of the family of Coccinellidae. Chromosomal diversity can avoided external morphological confusion and will be help full in correct identification of the species hence, present study was undertaken.

Materials and Methods

Materials used and Methodology adapted for biodiversity of Coccinellids and Chromosomal studies are given under separate chapter Materials and Methods.

Morphological Biodiversity of Coccinellids

Key to the Sub Families of Coccinillidae

1. Mandibles serrated or bifurcate, phytophagus, dull in colour, more or less larger species .... *Epilachninae*

2. Mandibles pointed and not serrated or bifurcated. Predatory beetles, bright or shining coloured. Small or large species .... *Coccinellinae*

Key to the Tribes of Sub Families Coccinellinae

1. Mandibles with bifid tip and with basal tooth mostly oval or hemispherical .... *Coccinellini*

2. Mandibles with multidenticulate apex. Not oval .... *Psychoborini*

Key to the Genera of Coccinelini

1. Black zigzag marking on yellow elytra. Body rounded ..... *Menochilus*

   No black zigzag lines on elytra

   Body subcircular, oval; hind tibia with spur, head black with two pale spots .... (2)

2. Scutellum triangular. Body subcircular.

   Antinnae longer than horns .... (3)

   Elytra base broder .... *Coelophora*

3. Body circular/rounded.. 3 *Menochilus*

    Body oval sub circular/hemispherical... 4

4. Hind tibia spurs indistinct, mandibles bifid ....*Harmonia* (5)

5. Antennae slightly longer than frons, ... *Coccinella*

    Mid tibial two spurs present, three and half spots
    on elytra .... *Coccinella*

    Hind tibial spurs two and small.......................................*Coelophora* (6)

6. Mandibles not multidentate,... *Vernia*

    Body hemispherical .... (7)

7. Mandibles multidentate

    Hind tibial spur absent, grey coloured ....*Illeis*

GENUS: *MENOCHILUS TIMBERLAKE*, 1943

The genus *Menochilus* is erected by Timberlake in 1943

Menochilus : Timberlake 1943, Hawaii. Plant. Rec. 47(1) 40.

Chilomenes : Mulsant 1850, Trim – spec. securipalp: 429

Chilomenes : Crotch, 1874. Rev. cocci. : 179 (pars).

Type species : *Coccinelee 6. maculate* Fabricius, 1781.

Under the genus *Menochilus* initially only single species *M. sexmacualtusi* was described by Fabricius in 1781. However, recently, Sathe and Bhosale (2001) added two species and Patil and Sathe (2003) added one more species of this genus thus, from India five species have been reported and described. The genus shows following characters: Body round slightly longer than wide and medium sized; prosternum with two carinae; anterior margin of the mesostrenum bisinuate and weakly emerginate in middle; elytral base broader, external margin of elytra not expanded but inclined; 1st abdominal sternum with an oblique lateral line; middle and hind tibia with two spurs. In the present text 9 new species are described.

1. *MENOCHILUS SANGLIENSIS* SP. NOV. (Plate 4; Figure 37)

Male

5.25 mm long and 4.10 mm wide, body medium sized and covex on dorsal side; pronotum 1.21mm long and 2.65 mm wide; antenna 0.85 mm long; elytra 4.52 mm long and with three prominent black zigzag spots, middle spot 'W' shaped; general body colour is light yellow.

Prey – Aphid, *Rhophalosiphum maidis* (Fitch)

Prey plant – Maize (*Zea mays*)

Locality – Palus, Dist. Sangli

Coll. – S. S. Patil from Feb. to March

Holotype – 17-2-2011, on card sheet, ASC Palus.

2. *MENOCHILUS SEXMACULATUS* (FAB) (Plate 4; Figures 38, 42, 43, 46)

Male

5.26 mm long and 4.12 mm wide, body round, medium sized; pronotum 1.23 mm long and 2.67 wide; antenna 0.86 long; elytra 4.54 mm long and 3.5 mm wide with three black spots; middle spots 'V' shaped with some extension, another elytra spot more or less 'C' shaped. Posterior spot semicircular, general body colouration light yellow, darker than *M. sangliensis*.

Prey – Delfacids, Aphids, *R. maidis*

Prey plant – Maize (*Z. mays*), Jowar (*Sorghum vulgare*)

Locality – Walwa, Dist. Sangli

Coll. – S. S. Patil from August to September.

3. *MENOCHILUS PATILENSIS* SP. NOV. (Plate 4, Figure 39)

Male

4.8 mm long and 4.67 mm wide, body round, medium sized; pronotum 1.00 mm long and 2.53 wide; antenna 0.74 long; elytra 4.44 mm long and 3.17 mm wide with three black spots; posterior black spots reduced, middle and anterior spots thin. General body colouration yellow darker than *M. sexmaculatus*.

Prey – Delfacids, Aphid, *Aphis craccivora* Koach.

Prey plant – Maize (*Z. mays*), Jowar (*S. vulgare*), Cotton (*Gossipium*)

Locality – Kolhapur, Dist. Sangli

Coll. – S. S. Patil from August to October.

Holotype – 15-VIII-2012, on card sheet, ASC Palus.

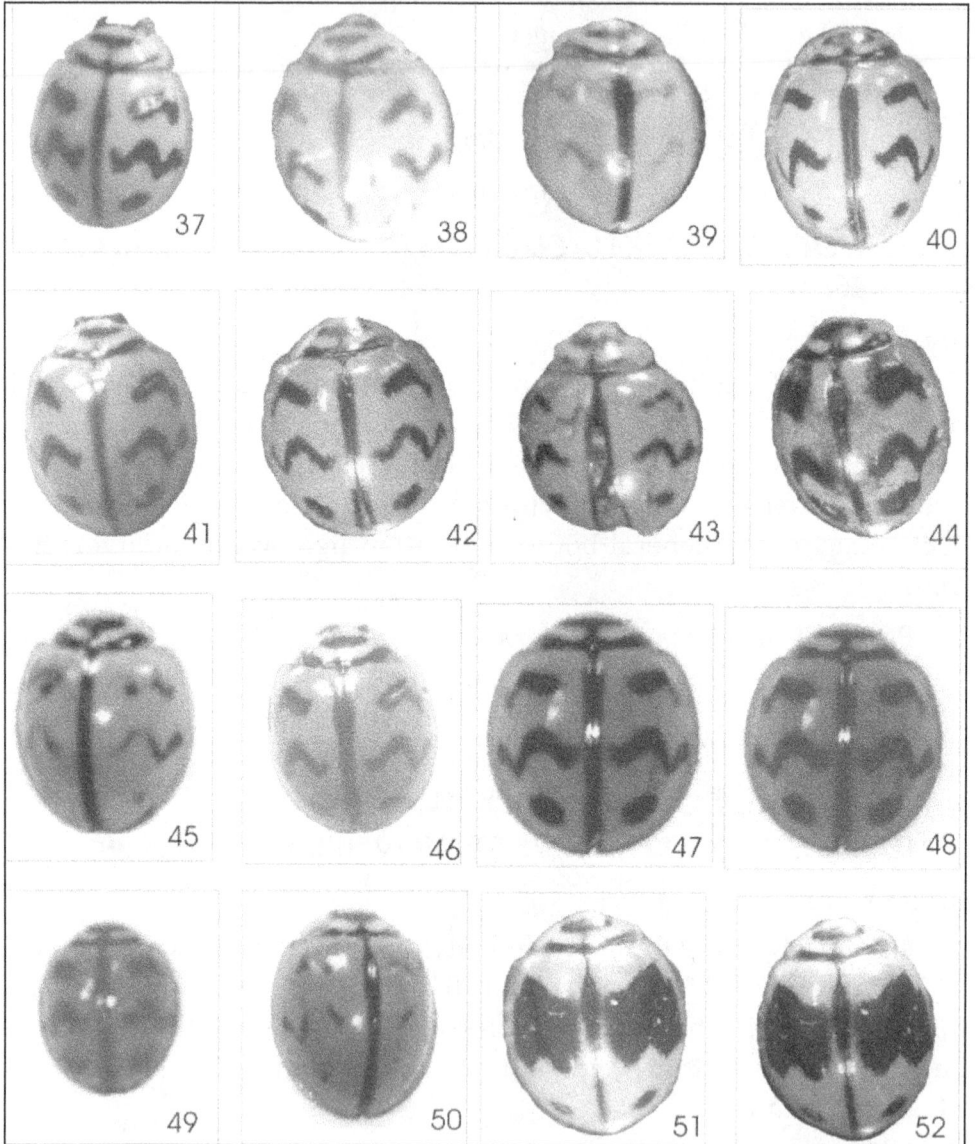

Plate 4

Figures 37–52.

4. *MENOCHILUS VAISHALI* P. and S. (Plate 4, Figures 40, 41)

Male

5.84 mm long and 4.4 mm wide, body round, medium sized; pronotum 1.12 mm long and 2.45 wide; antenna 0.75 long; elytra 5 mm long and 2.17 mm with three very prominent black spots; posterior is reduced to oval and anterior is broadly 'C' shaped. General body colouration light yellow darker than *M. patilensis* sp. nov.

Prey – Delfacids, Aphid, *R. maidis*

Prey plant – Maize (*Z. mays*)

Locality – Shirala Dist. Sangli

Coll. – S. S. Patil from November to January.

5. *MENOCHILUS MARATHI* SP. NOV. (Plate 4; Figure 44)

Male

5.82 mm long and 4.2 mm wide, body round, medium sized; pronotum 1.10 mm long and 2.43 wide; antenna 0.75 long; elytra 4 mm long and 2.15 mm wide with three very prominent black spots; anterior spot is very broad, more or less triangular, middle spot is broadly 'C' shaped, posterior spot is broad and elongated. General body colour is yellow orange and shape is oval.

Prey – Delfacids, Aphid, *R. maidis*

Prey plant – Maize (*Z. mays*), Jowar (*S. vulgare*)

Locality – Palus, Dist. Sangli

Collection – S.S. Patil from November to January.

Holotype – 20-XI-2012, on card sheet, ASC Palus.

6. *MENOCHILUS APHIDIVOURI* SP. NOV. (Plate 4; Figures 47, 48)

Male

5.44 mm long and 4.60 mm wide, body round, medium sized; pronotum 0.90 mm long and 2.55 wide; antenna 0.70 long; elytra 4.42 mm long and 3.15 mm wide with three black prominent spots; middle is 'W' shaped. General body colour is dark brown.

Prey – *A. craccivora*

Prey plant – Cotton (*Gossypium hirsutum*)

Locality – Khanapur, Dist. Sangli

Coll. – S. S. Patil from August to October.

Holotype – 18-VIII-2012, on card sheet, ASC Palus.

7. *MENOCHILUS* SP. NOV. (Plate 4; Figure 49)

Male

5.35 mm long and 4.50 mm wide, body round, medium sized; pronotum 0.70 mm long and 3.00 wide; antenna 0.90 long; elytra 4.80 mm long and 3.75 mm wide with three prominent black spots; middle is 'W' shaped. General body colour is brown.

Prey – Aphid, *R. maidis*

Prey plant – Maize (*Z. mays*)

Locality – Palus, Dist. Sangli

Coll. – S. S. Patil from March to May.

Holotype – 23-III-2011, on card sheet, ASC Palus.

8. *MENOCHILUS PALUSI* SP. NOV. (Plate 4; Figure 50)

Male

5.01 mm long and 4.22 mm wide, body round, medium sized; pronotum 0.50 mm long and 2.10 wide; antenna 0.50 long; elytra 4.20 mm long and 3.05 mm wide with three prominent black spots; spots are very much reduced, on the thorax two transverse black spots, general body colour brown, body oval and elongated.

Prey – Aphid, *R. maidis*

Prey plant – Maize (*Z. mays*)

Locality – Palus, Dist. Sangli

Coll. – S. S. Patil from March to May.

Holotype – 20-III-2012, on card sheet, ASC Palus.

9. *MENOCHILUS SATHEI* SP. NOV. (Plate 4; Figure 51)

Male

5.80 mm long and 4.03 mm wide, body round, medium sized; pronotum 1.13 mm long and 2.42 wide; antenna 0.70 long; elytra 5.1 mm long and 2.17 mm wide with three prominent black spots; General body colour is light yellow to grey colour. Elytra is with two spots, middle spot is very large black and heart shaped. Posterior spot is very small and transversally elongated.

Prey – Aphid, *R. maidis,* Delfacids

Prey plant – Jowar (*S. vulgare*)

Locality – Palus, Dist. Sangli

Coll. – S. S. Patil from June to August.

Holotype – 20-VI-2012, on card sheet, ASC Palus.

10. *MENOCHILUS INDICA* SP. NOV. (Plate 5; Figure 53)

Male

5.80 mm long and 4.03 mm wide, body round, medium sized; pronotum 1.13 mm long and 2.42 wide; antenna 0.70 long; elytra 5.1 mm long and 2.17 mm wide with three prominent black spots; General body colour is yellowish dark, two elytral black spot present; posterior largest and meeting on each elytra, forming half moon shape at anterior side, medium size black spot present. Body shape is rounded.

Prey – Aphid, *R. maidis*

Prey plant – Maize (*Z. mays*)

Locality – Palus, Dist. Sangli

Coll. – S. S. Patil from June to August.

Holotype – 28-VI-2012, on card sheet, ASC Palus.

11. *MENOCHILUS SHIVAJIANSIS* S. and B. (Plate 5; Figure 54)

Male

5.44 mm long and 4.60 mm wide, body round, medium sized; pronotum 0.90 mm long and 2.55 wide; antenna 0.70 long; elytra 4.42 mm long and 3.15 mm wide with three black prominent spots; General body colour light yellowish, two elytral spots present; posterior spot is inverted, 'C' shaped and posterior spot is very much large, occupying at least ¼ of the elytra, the same spot joins with the neighbouring elytra forming half moon shaped black spot, body rounded, elytral corner not angular posteriorly and elytra tips not acute.

Prey – *A. craccivora*

Prey plant – Cotton (*G. hirsutum*)

Locality – Khanapur, Dist. Sangli

Coll. – S. S. Patil from August to October.

GENUS: *HORMONIA* MULSANT, 1850

Mulsant (1850) erected the genus *Hormonia*. *Hormonia mulsant* 1850, *Spec. Trim, Securipalp*: 74,75. *Ptychantatis* Crotch, 1874, *Rev. Cocci.*, 122. *Callineda* Crotch, 1871. *Cat. Cocci.*, 6. *Leis.* Mulsant, 1850, *Spec. Trim. Securipalp.*, 241.

Six species namely *H. axyridis* (Pallus, 1773); *H. octomaculata* (Fabricius, 1981); *H. dimidiate* (Fabricius, 1981); *H. arcauta*(Fabricius, 1987); *H. sedecim-notata* (Fabricius, 1980) and *H. soyabinii* (Patil and Sathe, 2001) have been discovered.

The genus characterized by following characters: Body oval, longer than broad, moderately convex and medium sized; antenna longer than frons, 8th segment transverse and strongly broadening apically; lateral sides of pronotum strongly arcuated and marginated while basal corners rounded; pro-sternal process with or without two carinae; triangular, elytra yellow, orange or red with variable black pattern, middle tibia with spurs and post coxal line incomplete.

HORMONIA SOYABINII P. and S. (Plate 5; Figure 55)

Female

Body elongated oval, 6.45 mm long and 4.45 mm wide, weakly convex on dorsal side, pronotum 2.02 mm long and 4.25 wide, glabrous and without spot; antenna 1.60 mm long; elytra 4.25 mm long and 2.22 mm wide with four and half black spots; General body colour dark yellow to light brown, elytra with anterior small rounded spots and one large black spot meeting each other forming dumbbell shape. Prothorax without black spots. Anterio-lateral edges of elytra somewhat angular, elytral tips are pointed.

Prey – Aphid *R. maidis*

Prey plant – Maize (*Z. mays*)

Locality – Palus, Dist. Sangli

Coll. – S. S. Patil from February to April.

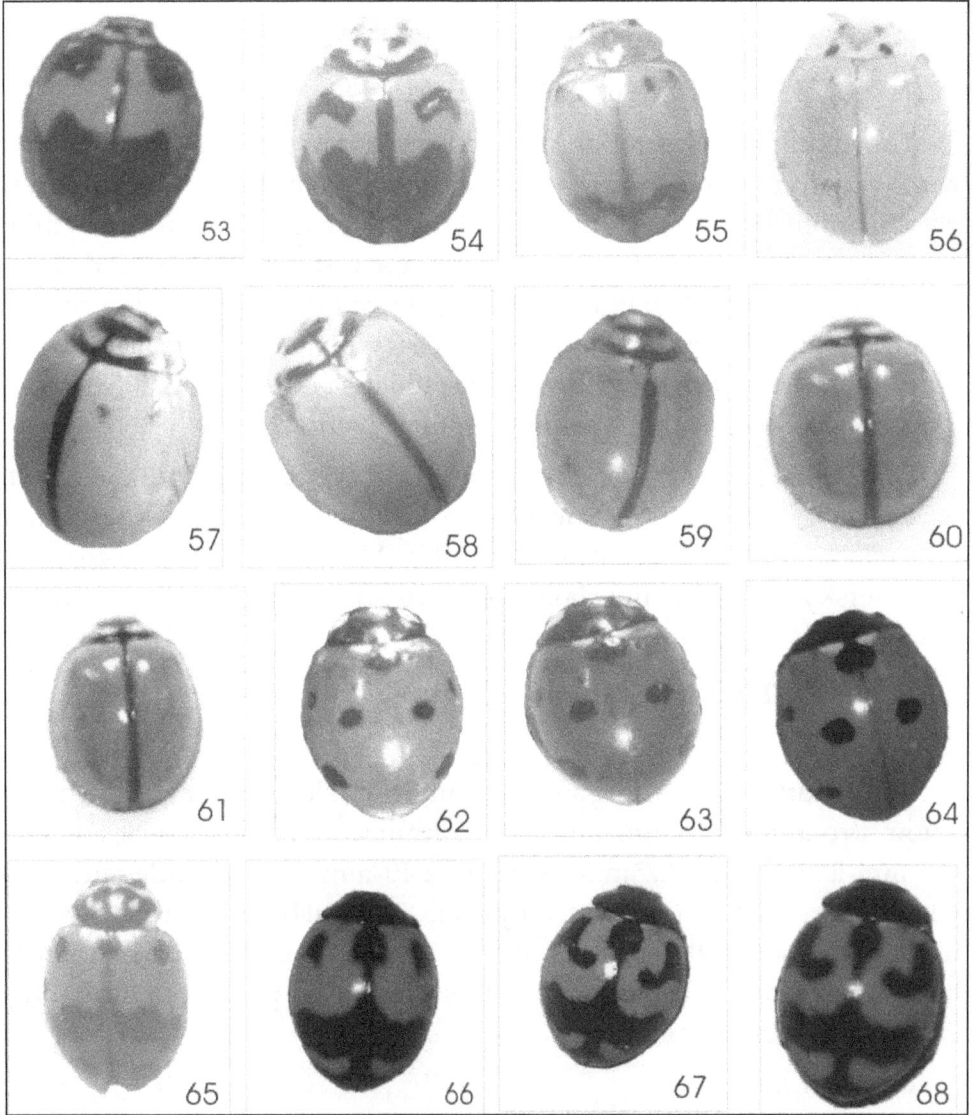

Plate 5

Figures 53–68.

GENUS: *ILLEIS* MULSANT, 1850

Mulsant (1850) proposed the genus *Illeis*

Illeis Mulsant 1850, Securipalp: 1026

Egleis Mulsant, 1850 *I. C.* : 167

Type species: *Coccinella cincta* Fabricius 1798.

The genus is characterized by body medium sized, short, oval and weakly convex; frons narrow; eyes with a shallow postantennal emergination; antennae distinctly longer than head width and very thin; mandible usually with multidenticulate apex, rarely with a bifid apex; terminal segment of maxillary palpus strongly transvers; scutellum transverse and triangular; prosternal process with parallel carinae; femoral line of 1st abdominal sternum incomplete; tibiae slender and simple without spurs; tarsal claws with a quadrate tooth at its base.

Under this genus 8 species have been described from India. *I. cincta, I. bistigmosa, I. indica, I. cincta amamiana. I. sathei* S and B, *I. darbari* S and B, *I. chilliei* P and S, *I. gavari, I. therioaphis* (Miyatake, 1959, Crotch 1874, Korschefsky 19323, Timberlake 1943, Bielawski 1961) are prominent species.

1. *ILLEIS BISTIGMOSA* (Plate 5; Figure 56)

Male

Body short, 4.80 mm long and 3.25 mm wide, slightly elongated whitish yellow, pronotum 1.03 mm long and 2.30 mm wide without spots, antenna 1.35 mm long, elytra 3.75mm long and 2.40 mm wide, middle and apex of lateral lobes of tegmen covered by setae. General body colour yellowish. Elytra with one small black spot at anterior and no black spots on middle or posterior side. Two black oval spots present on prothorax. Prothorax and head lighter in colour.

Prey – *A. craccivora*

Prey plant – Sauf (*Foenicaluns vulgare* Mill)

Locality – Shirala, Dist. Sangli

Coll. – S. S. Patil from November to February.

GENUS: *VERNIA* MULSANT, 1850

Mulsant erected the genus *Vernia* in 1850

Coccinella discolor Fabricius, 1798, Suppl. Ent. Syst. : 77

Vernia discolor : Mulsant, 1850, Spec. Trim. Securipalp: 369

Type species: *Vernia discolor*.

The genus shows following features : body hemispherical, glabrous, dorsum orange brown, frons sometimes with black spot, pronotum with discal spots, scutellum black, elytron with very narrow black sutural area, femora brown to black, tibiae and tarsae brown, prosternal carinae reaching half of prosternal length.

From India 5 species have been reported by Sathe and Bhosale (2001). *V. discolor, V. kiotoensis, V. cardoni* Omkar, *V. polyphagae, V. vulgeri* (Sathe and Bhosale, 2001; Patil and Sathe, 2003) have been reported. *V. cardoni* have been reported Uttar Pradesh.

1. *VERNIA INDICA* SP. NOV. (Plate 5; Figures 57, 58)

Female

Body oval, 5.5 mm long and 4.82 mm wide, dorsum convex, under side black, pronotum 1.07 mm long and 2.95 wide, antenna 0.57 mm long; elytra yellowish brown, 4.77 mm long and 2.27 mm wide. General body colour dark yellowish grey. No elytral spots present. On pro-thorax two transverse black spots present.

Prey – Aphid *R. maidis*

Prey plant – Hybrid Jowar (*S. vulgare*)

Locality – Walwa, Dist. Sangli

Coll. – S. S. Patil from January to April.

Holotype – 10-II-2012, on card sheet, ASC Palus.

2. *VERNIA POLYPHAGAE* S. and B. (Plate 5; Figures 60, 61)

Female

Body medium sized, 4.00 mm long and 3.89 mm wide, oval, dorsum convex, under side black, pronotum 1.04 mm long and 2.91 wide, antenna 0.96 mm long; elytra redish brown 3.20 mm long and 1.85 mm wide, siphon with deep notch between lobes of siphonal capsul. General body colour light brown to dark brown. No elytral spots present. transverse two black spots and one yellow spot present on prothorax.

Prey – Aphid *R. maidis*

Prey plant – Hybrid Jowar (*S. vulgare*), CSH-5

Locality – Miraj, Dist. Sangli

Coll. – S. S. Patil from January to April.

GENUS: *HIPPODAMIA*

This genus shows following characters.

Small and oval body, elytra with orange red colour with six black spots on each; pronotum black with white border, antennae clubbed, tarsal segment three. This Genus is widely studied from USA. Under this genus reported species refer to *H. variegata* and *H. convergens*, *H. indica*. First recorded from India.

1. *HIPPODAMINA INDICA* SP. NOV. (Plate 5; Figure 65)

Female

Body large, 5.57 mm long and 4.40 mm wide, oval, strongly convex on dorsal side, pronotum black, 1.68 mm long, glabrous, elytra yellowish brown, 4.74 mm long. General body colour yellowish to orange, three half black spot present on elytra, middle spot is broad and zigzag shaped and meeting at mid point to the anterior elytra. One rounded spot present anteriorly and one rounded spot at mid portion of anterior side. Joining each elytral tip typically pointed, prothoracic spot with yellow marking.

Prey – Aphid *R. maidis*

Prey plant – Maize (*Z. mays*)

Locality – Tasgoan, Dist. Sangli

Coll. – S. S. Patil from January to April.

Holotype – 30-I-2012, on card sheet, A.S.C. Palus.

GENUS: *COCCINELLA* LENNAUES 1958

In 1758 Linnaeus erected the genus *Coccinella* and was primarily holarctic (Dobzansky, 1931). Casay (1899, 1908) studied the North American species *Coccinella*. Leng (1903). Dobzansky, (1931) reported 11 species and many subspecies. Brown (1962) reported 12 species. The species described by other workers under this genus are *C. novemnotata* (Herbst, 1793), *C. transversalis* (Fabricius, 1781), *C. nigrita*, *C. septempunctata*, *C. repanda* (Thunberg, 1781), *C. dimidia* (Hope, 1831), *C. basalis* (Redtenbacher, 1848), *C. ainu* (Lewis, 1896), *C. explaneta*, *C. haseqawao* (Miyatake,1963) *C. quinquepunctata* (Linnaeus), *C. transversoquttata quinquepunctata* (Felderman), *C. trifasciata perflexa* (Mulsant) and *C. transversoquttata richardsoni* (Brown) from the peatch orchards of Canada from Ukraine D'yadechke (1954) reported *C. septempunctata*, to be the typical species of cereal fields; from India, Kapur, 1963 reported four species, *C. tebentena*, *C. lama*, *C. nigrovittata*, and *C. transversaslis* (Fabricius).

The genus Coccinella is confirmed by following characters. Body broadly oval, convex, head black with two pale spots on frons, antenne slightly longer than frons, post antennal process narrow and long, mesosternum flat to the anterior margin, scutelum triangular, elytra yellow, orange or red with variable black pattern, middle tibia with spurs and post coxal line incomplete.

1. *COCCINELLA SEPTEMPUNCTATA* THUNBERG, 1781 (Plate 5; Figures 62, 63, 64)

Female

Body small sized, 3.90 mm long and 3.70 mm wide, oval, pronotum 1.01 mm long and 2.87 wide, antenna 0.95 mm long; elytra redish brown 3.10 mm long and 1.75 mm wide. Body typically narrow towards posterior and anterior side. Head and prothorax black. Seven black spots are present on elytra, out of which one is common and situated at mid anterior. General body colour is redish brown, elytral tips are pointed.

Prey – Aphid *R. maidis*

Prey plant – Maize (*Z. mays*)

Locality – Tasgoan, Dist. Sangli

Coll. – S. S. Patil from January to April.

2. *COCCINELLA TABACI* P. and S. (Plate 5; Figure 66)

Body large, 6.3 mm long and 4.60 mm wide, oval, strongly convex on dorsal side, pronotum black, 1.55 mm long, glabrous, elytra dark yellowish brown, 5.1 mm long and 4.1 mm wide with three and half black spots, antenna clavate, 0.60 mm long, anterior spot is more or less triangular, one black common spot at mid anterior portion. Posterior black spot joins with another elytra and from trishul shaped appearance. Prothorax and head black. Body shape elongated oval.

Prey – Aphid *Myzus persicae* (Satz)

Prey plant – Tobacco (*N. tabacum*)

Locality – Miraj, Dist. Sangli

Coll. – S. S. Patil from November to February

3. *COCCINELLA TRANSVERSALIS* (FAB.) (Plate 5, Figures 67, 68)

Body large, 5.75mm long, and 4.40mm wide, oval, strongly convex on dorsal side; pronotum black 1.68mm long, glaborous; elytra yellowish brown, 4.74mm long with three half black spots, ventral side black, antenna elevate, long brown.

Prey – Aphid *M. persicae*

Prey plant – Jawar (*S. vulgare*)

Locality – Khanapur, Dist. Sangli

Coll. – S. S. Patil from November to February.

6

Chromosomal Diversity of Coccinellids

1. *Menochilus vaishali* P. and S.

The results represented in Plate 6 and Figures 69 to 73 shows that in *Menochilus vaishali* the diploid number of 12 chromosomes was depicted by the spermatogonial metaphase. The spermatogenic meiotic metaphase – I was depicted by haploid number of six chromosomes. The karyotype shows, five pair of autosomes and X and Y sex chromosomes. Autosome number 1, 2, 4, 5, 6 and X chromosomes were submetacentric and autosome number 3, 7, 8, 9, 10 and Y chromosomes were metacentric.

2. *Menochilus sexmaculatus* (Fab.)

The results represented in Plate 7 and Figures 74 to 78 shows that in *Menochilus sexmaculatus* the diploid number of 12 chromosomes was depicted by the spermatogonial metaphase. The spermatogenic meiotic metaphase – I was depicted by haploid number of six chromosomes. The karyotype shows, five pair of autosomes and X and Y sex chromosomes. Autosome number 1 to 8, 10 and Y chromosomes were metacentric and autosome number 9 and X chromosomes were submetacentric.

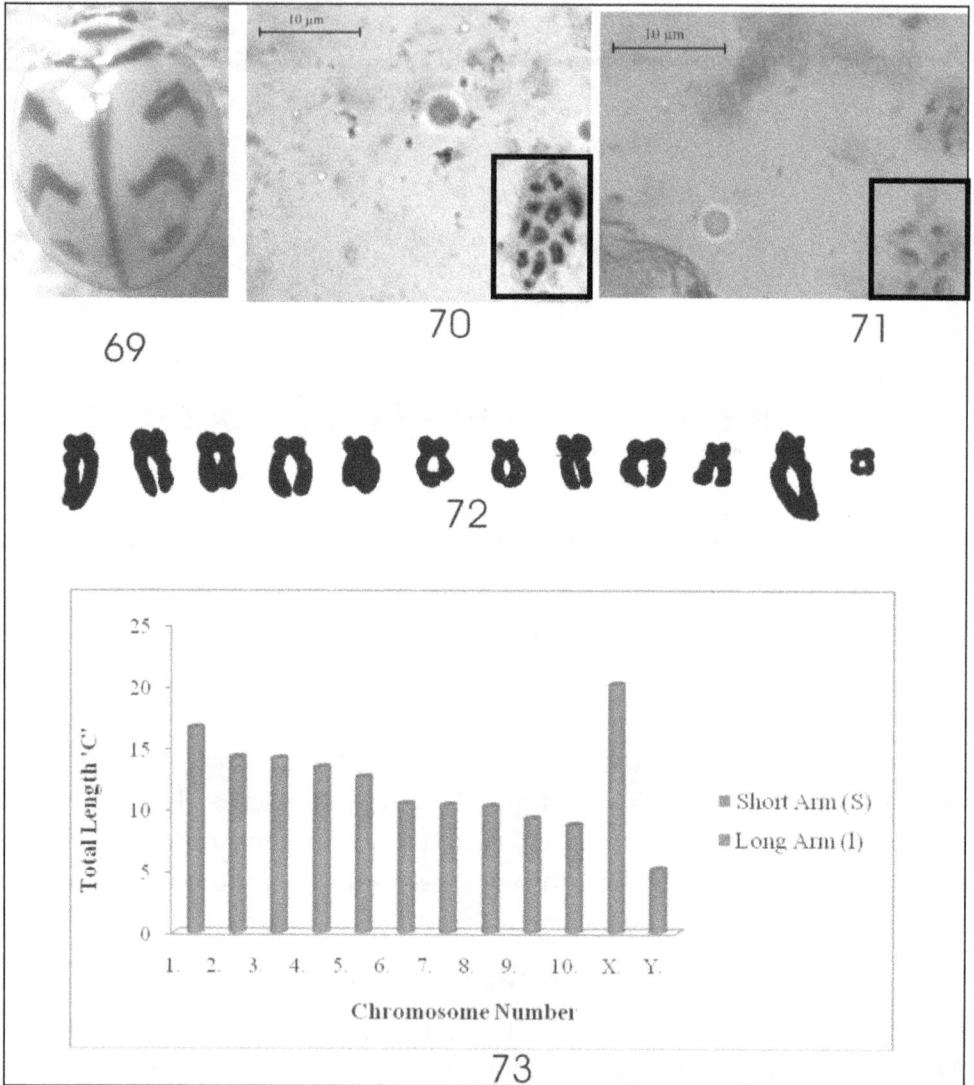

Plate 6

Figure 69: *Menochilus vaishali*; **Figure 70**: Spermatogonial mitotic metaphase; **Figure 71**: Spermatogenic meiotic metaphase-I; **Figure 72**: Karyograph; **Figure 73**: Ideograph.

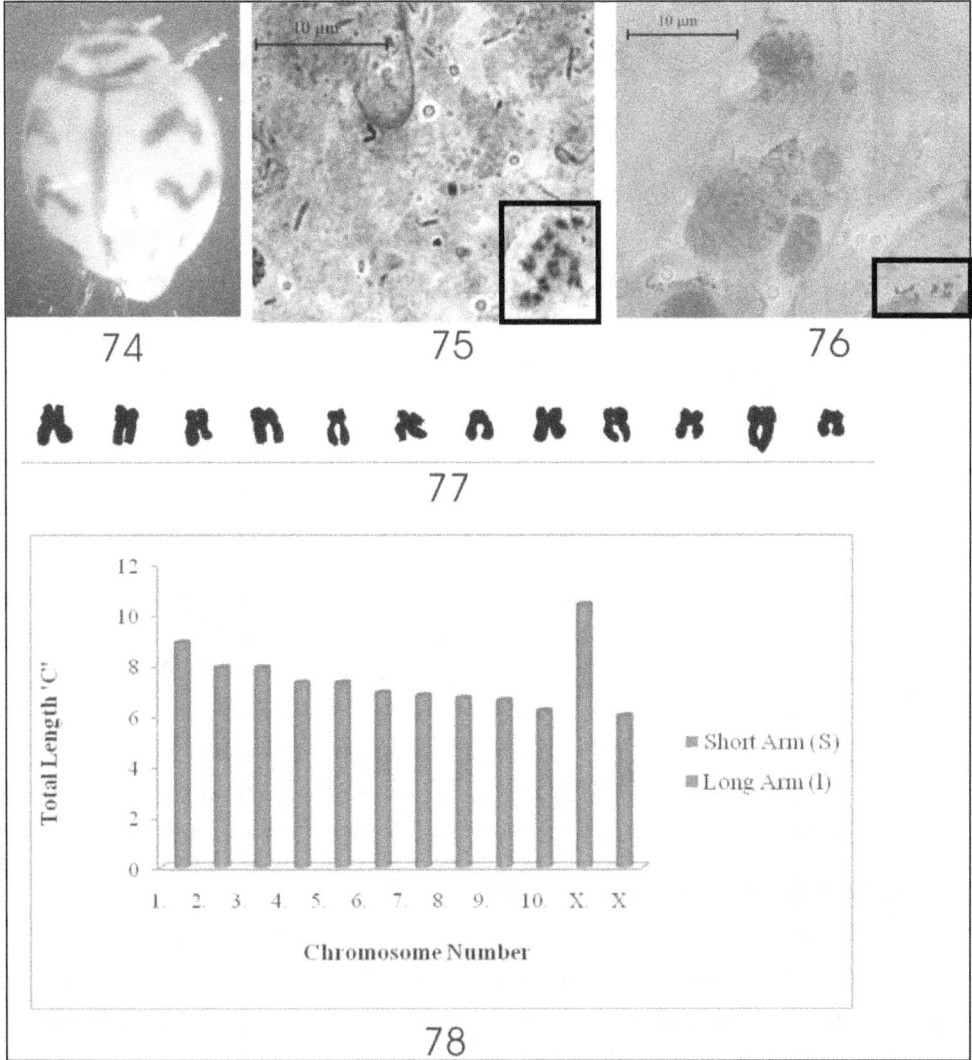

Plate 7

Figure 74: *Menochilus sexmaculatus*; **Figure 75**: Spermatogonial mitotic metaphase; **Figure 76**: Spermatogenic meiotic metaphase-I; **Figure 77**: Karyograph; **Figure 78**: Ideograph.

3. *Menochilus sangliensis* sp. nov.

The results represented in Plate 8 and Figures 79 to 84 shows that in *Menochilus sangliensis* sp. nov. the diploid number of 12 chromosomes was depicted by the spermatogonial metaphase. The spermatogenic meiotic metaphase – I was depicted by haploid number of six chromosomes. The karyotype shows, five pair of autosomes and X and Y sex chromosomes. All the autosome pairs and X and Y sex chromosomes were metacentric.

4. *Hormonia maharashtri* sp. nov.

The results represented in Plate 9 and Figures 85 to 89 shows that in *Hormonia maharashtri* sp. nov. the diploid number of 14 chromosomes was depicted by the spermatogonial metaphase and the haploid number of 7 chromosomes was depicted by spermatogenic meiotic metaphase – I. The karyotype shows, six pair of autosomes and X and Y sex chromosomes. The autosome number 1, 2, 3, 6 to 12 and X and Y sex chromosomes were metacentric while, autosome number 4 and 5 were submetacentric.

5. *Menochilus aphidivouri* sp. nov.

The results represented in Plate 10 and Figures 90 to 94 shows that, in *Menochilus aphidivouri* sp. nov. the diploid number of 12 chromosomes was depicted by the spermatogonial metaphase and the haploid number of 6 chromosomes was depicted by spermatogenic meiotic metaphase – I. The karyotype shows, five pair of autosomes and X and Y sex chromosomes. The autosome number 1, 2, 3, 5, 7, 8, and X and Y sex chromosomes were metacentric while, autosome number 4, 6, 9 and 10 were submetacentric.

6. *Menochilus patilensis* sp. nov.

The results represented in Plate 11 and Figures 95 to 99 shows that in *Menochilus patilensis* sp. nov. the diploid number of 12 chromosomes was depicted by the spermatogonial metaphase and the haploid number of 6 chromosomes was depicted by spermatogenic meiotic metaphase – I. The karyotype shows, five pair of autosomes and X and Y sex chromosomes. All the autosome pairs as well as X and Y sex chromosomes were metacentric.

7. *Menochilus palusi* sp. nov.

The results represented in Plate 12 and Figures 100 to 104 shows that in *Menochilus palusi* sp. nov. the diploid number of 12 chromosomes was

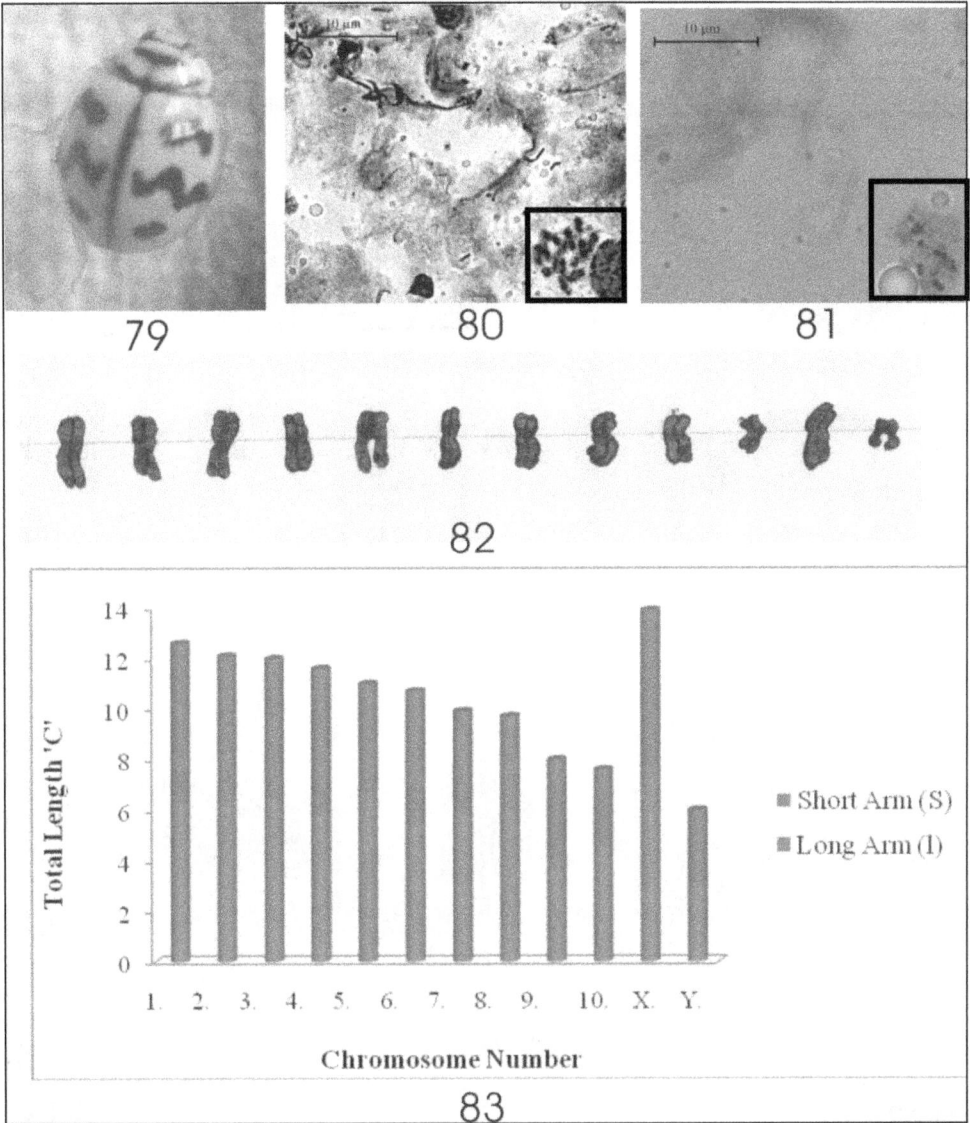

Plate 8

Figure 79: *Menochilus sangliensis*; **Figure 80**: Spermatogonial mitotic metaphase; **Figure 81**: Spermatogenic meiotic metaphase-I; **Figure 82**: Karyograph; **Figure 83**: Ideograph.

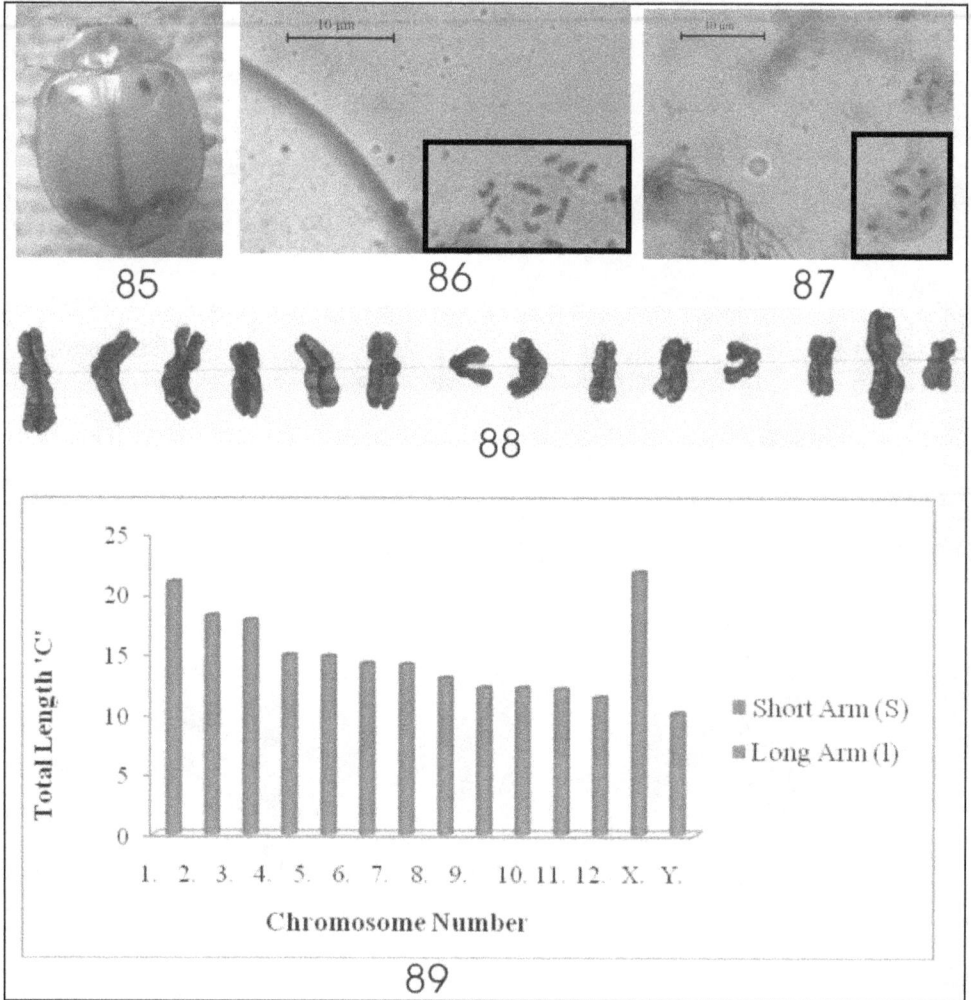

Plate 9

Figure 85: *Harmonia maharashtri*; **Figure 86**: Spermatogonial mitotic metaphase; **Figure 87**: Spermatogenic meiotic metaphase-I; **Figure 88**: Karyograph; **Figure 89**: Ideograph.

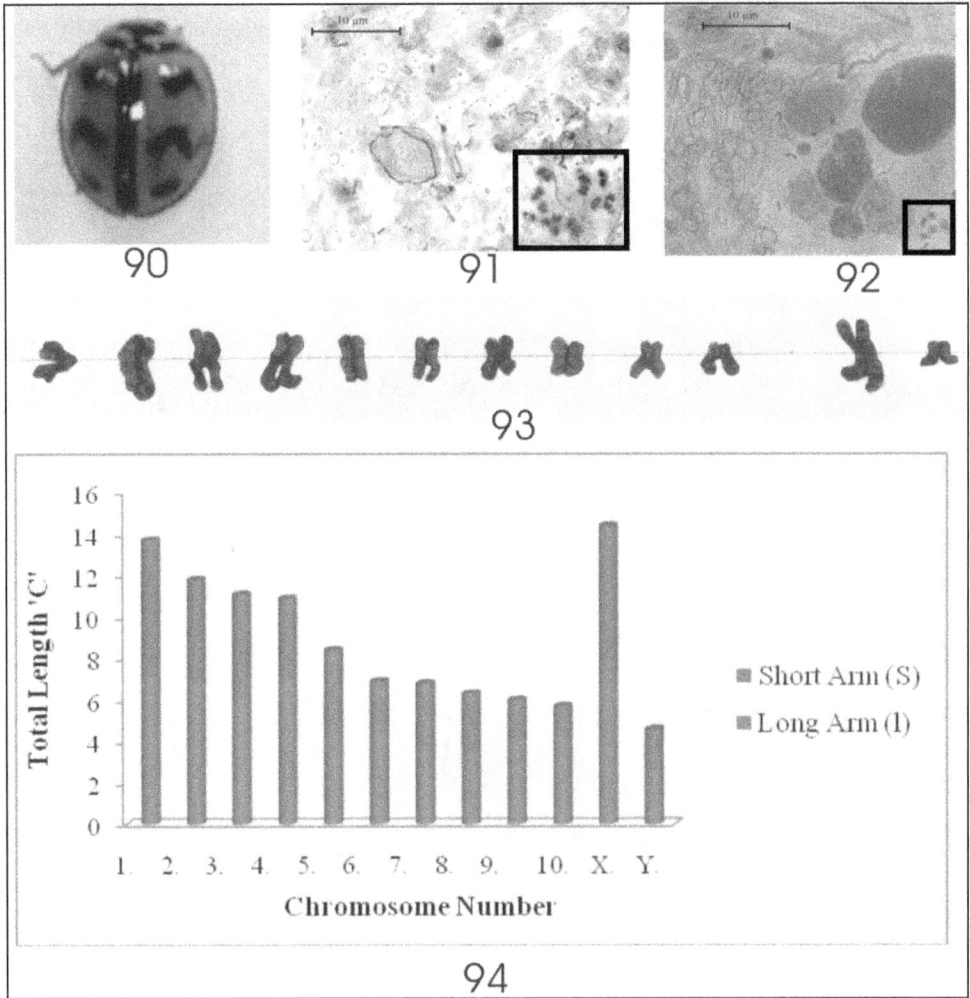

Plate 10

Figure 90: *Menochilus aphidivouri*; **Figure 91**: Spermatogonial mitotic metaphase; **Figure 92**: Spermatogenic meiotic metaphase-I; **Figure 93**: Karyograph; **Figure 94**: Ideograph.

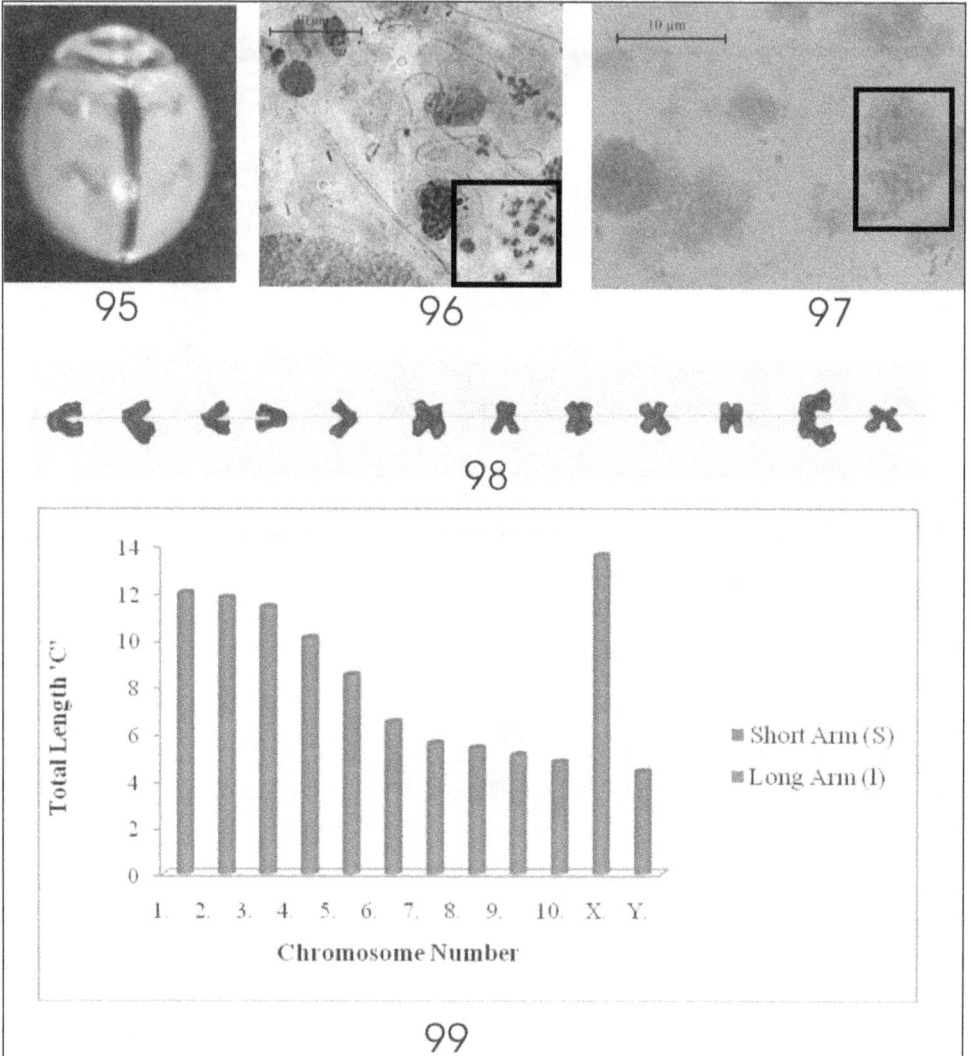

Plate 11

Figure 95: *Menochiius patilensis;* **Figure 96**: Spermatogonial mitotic metaphase;
Figure 97: Spermatogenic meiotic metaphase-I; **Figure 98**: Karyograph;
Figure 99: Ideograph.

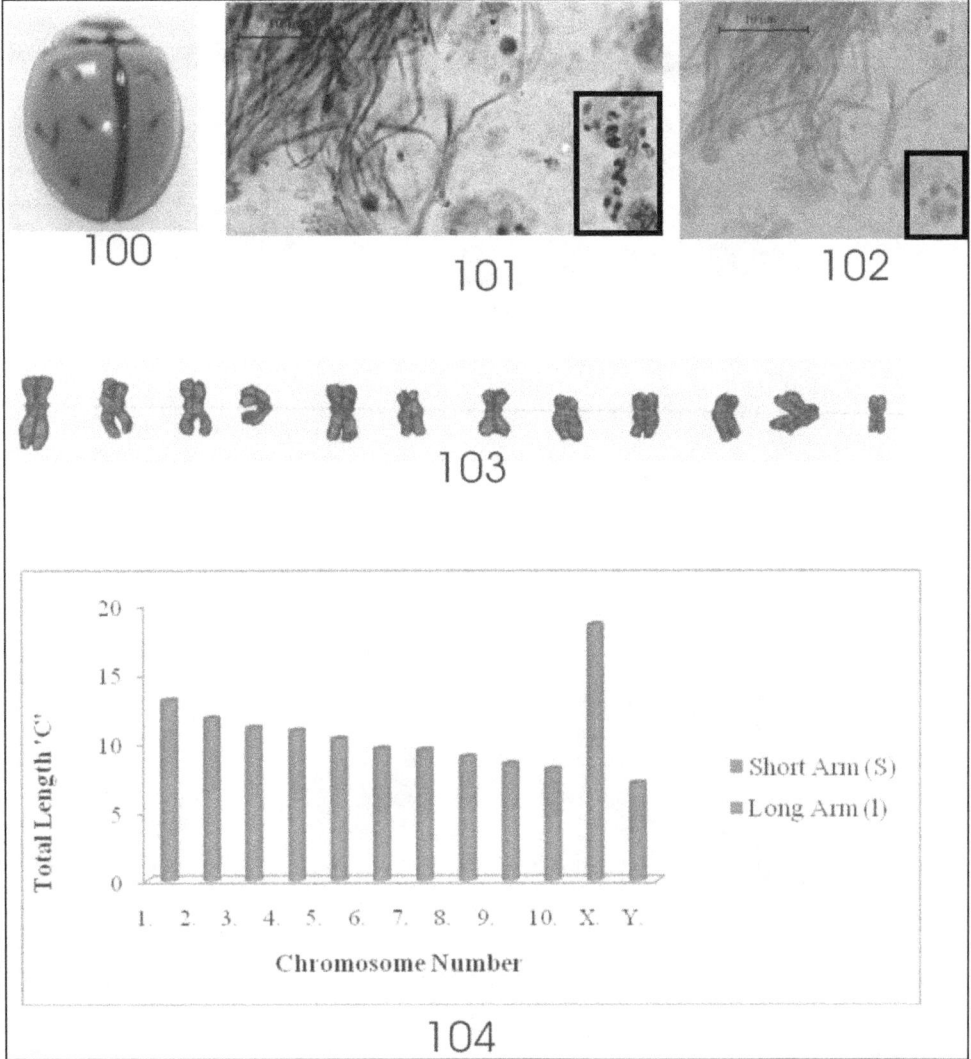

Plate 12

Figure 100: *Menochilus polusi*; **Figure 101**: Spermatogonial mitotic metaphase; **Figure 102**: Spermatogenic meiotic metaphase-I; **Figure 103**: Karyograph; **Figure 104**: Ideograph.

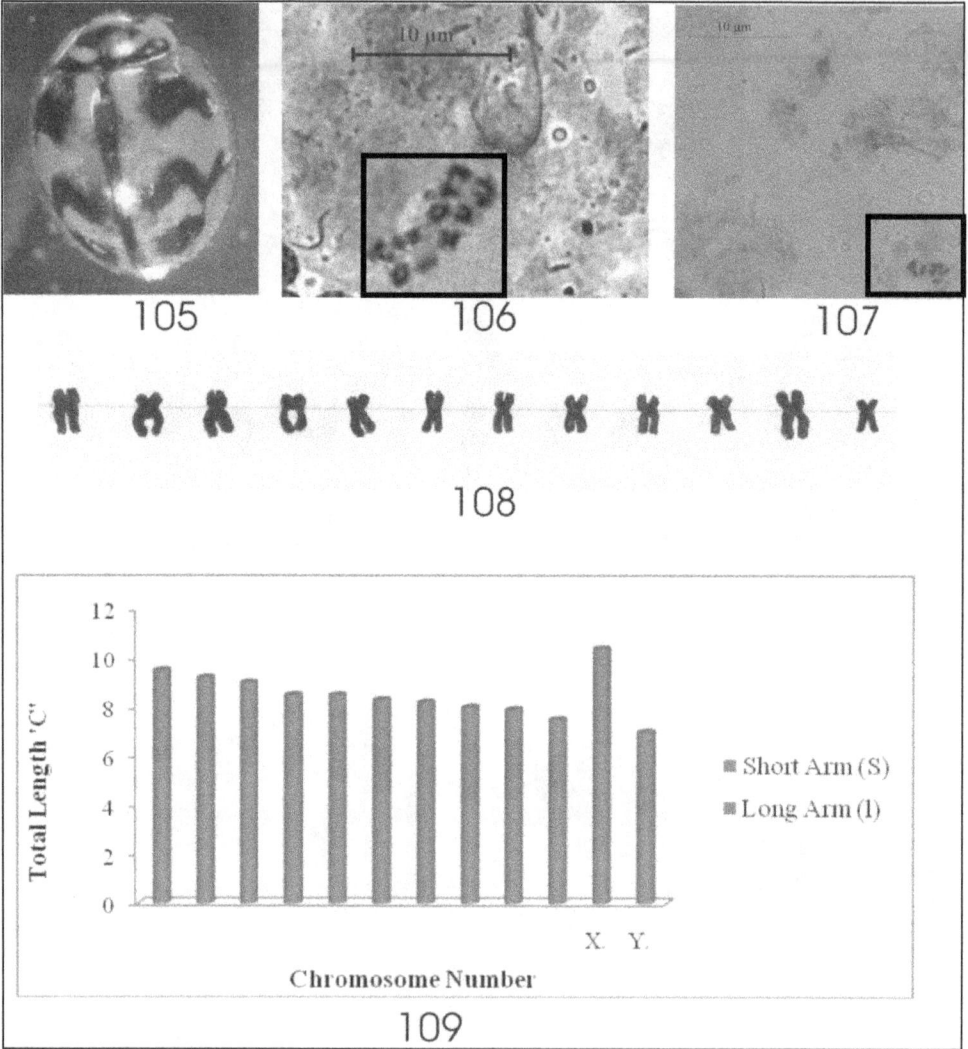

Plate 13

Figure 105: *Menochilus marathi*; **Figure 106**: Spermatogonial mitotic metaphase; **Figure 107**: Spermatogenic meiotic metaphase-I; **Figure 108**: Karyograph; **Figure 109**: Ideograph.

Table 1: Morphometric Characteristic of the Chromosomes of *Menochilus vaishali*.

Chromosome No.	Long Arm (l)	Short Arm (S)	Total Length (C = l + S)	'd' Value (l – S)	'r' Value (l/S)	'i' Value (S/C X 100)	Centromeric Position
1.	11.4	5.3	16.7	6.1	2.1509	31.7365	SM
2.	9.1	5.2	14.3	3.9	1.7500	36.3636	SM
3.	8.3	5.9	14.2	2.4	1.4067	41.5492	M
4.	9.4	4.1	13.5	5.3	2.2926	30.3703	SM
5.	8.6	4.1	12.7	4.5	2.0975	32.2834	SM
6.	6.6	3.9	10.5	2.1	0.1946	37.1428	SM
7.	5.8	4.6	10.4	1.2	0.1739	44.2307	M
8.	5.9	4.4	10.3	1.5	1.3409	42.7184	M
9.	5.8	3.5	9.3	2.25	1.6571	37.6344	M
10.	5.2	3.6	8.8	1.6	1.4444	40.9090	M
X.	13.4	6.8	20.2	6.6	1.9705	33.6633	SM
Y.	2.9	2.3	5.2	0.6	1.2608	44.2307	M

Table 2: Morphometric Characteristic of the Chromosomes of *Menochilus sexmaculatus*.

Chromosome No.	Long Arm (l)	Short Arm (S)	Total Length (C = l + S)	'd' Value (l − S)	'r' Value (l/S)	'i' Value (S/C X 100)	Centromeric Position
1.	4.9	4.0	8.9	0.9	1.2250	44.9438	M
2.	4.2	3.7	7.9	0.5	1.1351	46.8354	M
3.	4.2	3.7	7.9	0.5	1.1351	46.8354	M
4.	4.2	3.1	7.3	1.1	1.3548	42.4657	M
5.	4.3	3.0	7.3	0.3	0.4109	41.0958	M
6.	3.6	3.3	6.9	0.3	0.4782	47.8260	M
7.	3.7	3.1	6.8	0.6	0.4558	45.5882	M
8.	3.5	3.2	6.7	0.3	0.4776	47.7611	M
9.	4.3	2.3	6.6	2.0	1.8695	34.8484	SM
10.	3.1	3.1	6.2	1.0	0.5000	50.0000	M
X.	6.6	3.8	10.4	2.8	1.7368	36.5384	SM
Y	3.2	2.8	6.0	0.4	0.4666	46.6666	M

Table 3: Morphometric Characteristic of the Chromosomes of *Menochilus sangliensis* sp. nov.

Chromosome No.	Long Arm (l)	Short Arm (S)	Total Length (C = l + S)	'd' Value (l − S)	'r' Value (l/S)	'i' Value (S/C X 100)	Centromeric Position
1.	6.3	6.3	12.6	0.0	1.0000	50.0000	M
2.	6.4	5.7	12.1	0.7	1.1228	47.1074	M
3.	6.4	5.6	12.0	0.8	1.1428	46.6666	M
4.	5.8	5.8	11.6	0.0	1.0000	50.0000	M
5.	6.7	4.3	11.0	2.4	1.5581	39.0909	M
6.	5.4	5.3	10.7	0.1	1.0188	49.5327	M
7.	5.9	4.0	9.9	1.9	1.4750	40.4040	M
8.	6.0	3.7	9.7	2.3	1.6216	38.1443	M
9.	4.0	4.0	8.0	0.0	1.0000	50.0000	M
10.	4.5	3.1	7.6	1.4	1.4516	40.7894	M
X.	7.3	6.6	13.9	0.7	1.1060	47.4820	M
Y.	3.1	2.9	6.0	0.2	1.0689	48.3333	M

Table 4: Morphometric Characteristic of the Chromosomes of *Hormonia maharashtri* sp. nov.

Chromosome No.	Long Arm (l)	Short Arm (S)	Total Length (C = l + S)	'd' Value (l − S)	'r' Value (l/S)	'i' Value (S/C X 100)	Centromeric Position
1.	12.1	8.9	21.0	3.2	1.3595	42.3809	M
2.	9.1	9.1	18.2	0.0	1.000	50.0000	M
3.	9.1	8.7	17.8	0.4	1.0459	48.8764	M
4.	9.6	5.3	14.9	4.3	1.8113	35.5704	SM
5.	9.5	5.3	14.8	4.2	1.7924	35.8108	SM
6.	7.1	7.1	14.2	0.0	1.0000	50.0000	M
7.	8.1	6.0	14.1	2.1	1.3500	42.5531	M
8.	7.1	5.9	13.0	1.2	0.4538	45.3846	M
9.	7.4	4.8	12.2	2.6	1.5416	39.3442	M
10.	6.8	5.4	12.2	1.4	1.2592	44.2622	M
11.	6.5	5.6	12.1	0.9	1.1607	46.2809	M
12.	6.3	5.1	11.4	1.2	1.2352	44.7368	M
X.	11.3	10.5	21.8	0.8	1.0761	48.1651	M
Y.	6.0	4.1	10.1	1.9	1.4634	40.5940	M

Table 5: Morphometric Characteristic of the Chromosomes of *Menochilus aphidivouri* sp. nov.

Chromosome No.	Long Arm (l)	Short Arm (S)	Total Length (C = l + S)	'd' Value (l – S)	'r' Value (l/S)	'i' Value (S/C X 100)	Centromeric Position
1.	7.1	6.6	13.7	0.5	1.0757	48.1751	M
2.	6.9	4.9	11.8	2.0	1.4081	41.5254	M
3.	6.1	5.0	11.1	1.1	1.2200	45.0450	M
4.	6.9	4.0	10.9	2.9	1.7250	36.6972	SM
5.	4.4	4.0	8.4	0.4	1.1000	47.6190	M
6.	4.5	2.4	6.9	2.1	1.8750	34.7826	SM
7.	3.7	3.1	6.8	0.6	1.1935	45.5882	M
8.	3.5	2.8	6.3	0.7	1.2500	44.4444	M
9.	4.0	2.0	6.0	2.0	2.0000	33.3333	SM
10.	4.2	1.5	5.7	2.7	2.8000	26.3157	SM
X.	7.3	7.1	14.4	0.2	1.0281	49.3055	M
Y.	2.7	1.9	4.6	0.8	1.4210	40.7725	M

Table 6: Morphometric Characteristic of the Chromosomes of *Menochilus patilensis* sp. nov.

Chromosome No.	Long Arm (l)	Short Arm (S)	Total Length (C = l + S)	'd' Value (l – S)	'r' Value (l/S)	'i' Value (S/C X 100)	Centromeric Position
1.	6.0	6.0	12.0	0.0	1.0000	50.0000	M
2.	6.7	5.1	11.8	1.6	1.3137	43.2203	M
3.	5.9	5.5	11.4	0.4	1.0727	48.2456	M
4.	5.4	4.7	10.3	0.7	1.1489	45.6310	M
5.	4.7	3.8	8.5	0.9	1.2368	44.7059	M
6.	3.5	3.0	6.5	0.5	1.1666	46.1538	M
7.	3.2	2.4	5.6	0.8	1.3333	42.8571	M
8.	3.0	2.4	5.4	0.6	1.2500	44.4444	M
9.	2.8	2.3	5.8	0.5	1.2173	45.0980	M
10.	2.4	2.4	4.8	0.0	1.0000	50.000	M
X.	6.8	6.8	13.6	0.0	1.0000	50.000	M
Y.	2.7	1.7	4.4	1.0	1.5882	38.6363	M

Table 7: Morphometric Characteristic of the Chromosomes of *Menochilus palusi* sp. nov.

Chromosome No.	Long Arm (l)	Short Arm (S)	Total Length (C = l + S)	'd' Value (l – S)	'r' Value (l/S)	'i' Value (S/C X 100)	Centromeric Position
1.	7.6	5.5	13.1	2.1	1.3818	41.9847	M
2.	6.4	5.4	11.6	1.2	1.2307	44.8275	M
3.	6.1	5.0	11.1	1.1	1.2200	45.0450	M
4.	5.8	5.1	10.9	0.7	1.1372	46.7889	M
5.	5.2	5.1	10.3	0.1	1.0196	49.5145	M
6.	4.8	4.8	9.6	0.0	1.0000	50.0000	M
7.	4.8	4.7	9.5	0.1	1.0212	49.4736	M
8.	5.8	3.2	9.0	2.6	1.8145	35.5555	SM
9.	5.0	3.5	8.5	1.5	1.4285	41.1764	M
10.	4.7	3.4	8.1	1.3	1.3827	41.9753	M
X.	9.5	9.1	18.6	0.4	1.0439	48.9247	M
Y.	3.6	3.5	7.1	0.1	1.0285	49.2957	M

Table 8: Morphometric Characteristic of the Chromosomes of *Menochilus marathi* sp. nov.

Chromosome No.	Long Arm (l)	Short Arm (S)	Total Length (C = l + S)	'd' Value (l − S)	'r' Value (l/S)	'i' Value (S/C X 100)	Centromeric Position
1.	5.8	3.7	9.5	2.1	1.5675	38.9473	M
2.	6.1	3.1	9.2	3.0	1.9677	33.6955	SM
3.	5.6	3.4	9.0	2.2	1.6470	37.7777	M
4.	5.5	3.0	8.5	2.5	1.8333	35.2941	SM
5.	5.0	3.5	8.5	1.5	1.4285	41.1764	M
6.	4.5	3.8	8.3	0.7	1.1842	45.7831	M
7.	4.7	3.5	8.2	1.2	1.3428	42.6829	M
8.	4.2	3.8	8.0	1.5	1.1315	47.5000	M
9.	4.2	3.7	7.9	1.5	1.1351	46.8354	M
10.	4.5	3.0	7.5	1.5	1.5000	40.0000	M
X.	6.4	4.0	10.4	2.4	1.6000	38.4615	M
Y.	3.7	3.3	7.0	0.7	1.1212	47.1428	M

depicted by the spermatogonial metaphase and the haploid number of 6 chromosomes was depicted by spermatogenic meiotic metaphase – I. The karyotype shows five pair of autosomes and X and Y sex chromosomes. The autosome number 1, to 7, 9, 10 and X and Y sex chromosomes were metacentric while, autosome number 8 was submetacentric.

8. *Menochilus marathi* sp. nov.

The results represented in Plate 13 and Figures 105 to 109 shows that in *Menochilus marathi* sp. nov. the diploid number of 12 chromosomes was depicted by the spermatogonial metaphase and the haploid number of 6 chromosomes was depicted by spermatogenic meiotic metaphase – I. The karyotype shows five pairs of autosomes and X and Y sex chromosomes. The autosome number 1, 3, 5 to 10 and X and Y sex chromosomes were metacentric while, autosome number 2 and 4 were submetacentric.

7

Concepts of Chromosomal Diversity

Dange and Rathore (2010) studied chromosomes in four species of Scarabaeidae. In *Heliocopris bucephales* (Fabricius) the diploid number of 20 chromosomes was epicted by the spermatogonial metaphase. The karyotype showed 9 paired autosome and X and Y sex chromosomes. Autosome pair 1 was metacentric and Autosome pairs 2 – 9 and X and Y sex chromosomes were sub-metacentric. When arranged according to size the autosomes showed a gradual decrease in size. Sufficient number of cells at Metaphase I[st] depicted 9 rod shaped autosomal bivalents and the sex bivalent Xy. Due to reproduction division in metaphase IInd two types of chromosomes X and Y have been observed in *Gymnopleurus dejeani* Castelnau diploid number 20 chromosomes were revealed by the spermatogonial metaphase. The karyotype was composed of 9 pairs of autosomes and X and Y sex chromosomes. Autosome pairs 1 – 3 were metacentric, autosome pairs 4 – 8 were submetacentric, X chromosome was metacentric whereas Y chromosome was acrocentric. Metaphase I showed 9 rods or dumb-bell shaped autosomal bivalents and sex bivalents in the form of parachute. In Metaphase II Y chromosome was observed in *G. gemmatus* Harold the diploid number of 20 chromosomes was revealed by spermatogonial metaphase. The karyotype was composed of 9 pairs of autosomes and X and Y sex chromosomes. Autosome pairs 2, 4, 5, 7, 9 were metacentric while, Autosome pairs 1, 3 and 6 were submetacentric. X and Y sex chromosomes were metacentric. Metaphase I showed 9 rod shaped autosomal bivalents and sex bivalents in the form of parachute. In Metaphase II Y chromosome was noticed.

In G. *miliaris* (Fabricius) Dange and Rathore (2010) reported diploid number of 18 chromosomes depicted by the spermatogonial metaphase. The karyotype comprised 8 pairs of autosomes and X and Y sex chromosomes, all pairs of autosomes were metacentric and X chromosome was submetacentric including Y chromosome was also metacentric. Eight autosomal bivalents were dumb-bell and rod shaped. In Metaphase II X chromosome was observed.

Smith and Virkki (1978) cytogenetically analyzed around half of the species of the sub-family Scarabaeinae which presented variation in diploid number or in the sex determination mechanism, 2n = 20, Xyp considered primitive in the family Scarabaeidae and for the order Coleoptera. Scarabaeinae is the most karyotypically diverse subfamily of Scarabaeidae, with variation in diploid numbers and sex determining mechanisms (Bione *et al.*, 2005a,b, Carbal-de-Mello *et al.*, 2008, Colomba *et al.*, 1996, Martins, 1994, Vitturi *et al.*, 2003 and Yadav and Pilliae, 1979). The smallest diploid number 2n = 3 + neo Xy depicted in *Eurysternus caribaeus* (Carbal-de-Mello *et al.*, 2007) and highest diploid number 2n = 10 + X0 exhibited by *Copris fricator* (Joneja, 1960) which is comparable to *Apogonia* spp. (Saha, 1973) of Milolonthinae. In Scarabaeinae 143 are known cytologically out of these 7 species showed the 'Modal number' (Duff, 1970, Kacker 1970 and Lahiri and Manna, 1969). In *Gymnopleurus dejeani* and *G. gemmatus* exhibited 2n = 9 + Xyp Modal number of Scarabaeidae, where as confirmed an earlier (Manna and Lahiri, 1972) *Heliocopris* analyzed hereis, presented a karyotype with 2n = 9 + Xy, however *G. miliaris* showed a karyotype 2n = 18. The relatively large size of pair 1, which corresponded to the larges element of the complement, characterized a karyotypic asymmetry in these species.

The reduction of diploid number of 2n = 18 and the relatively large size of pair 1 when compare to the other chromosomes of the karyotype suggests the occurrence of a pericenric inversion followed by a fusion between autosomes from an ancestral karyotype with 2n = 20. Similarly, Bione *et al.* (2005a, b) observed rearrangements and represented the main karyotypic changes involved in the chromosomal evolution of scarabaeidae (Cabral-de-Mello *et al.*, 2007, 2008 and Yadav and Pilliae, 1979). According to Virkki, (1957) the chromosome number 2n = 21 (X0) in *Copris fricator* may have been evolved through Geotrupine karyotype 2n = 22 (Xyp). Since Geotrupinae is anatomically close to Coprinae than any other subfamily, these hypotheses get enough support. Baimai (1998) studies the *Drosophila melanogaster* group for mitotic karyotypes in *D. birchii* an endemic species of northern Australia and Papua New Guinea and *D. kikkawai*, sub cosmopolitan city recidal species. These species exhibited extensive variation in mitotic karyotypes which are expanded by heterochromatin additions in to the 4th chromosome (microchromosome) and sex chromosomes. A similar phenomenon of heterochromatin variation has been observed in *D. meridionalis* and *D. serido* of the *D. replete* group which are endemic to South America. None of these led to new species designations. However, detailed cytogentic studies of the Hawaiian drosophila, *D. bostrycha* and *D. disjuncta*, have led to a discovery of a new sibling species *D. affinidisjuncta*, alopatrically occurring in west Maui. Such presumed alopatric speciation of the Hawaiian drosophila correlates with heterochromatin accumulation in autosomes as well as in sex chromosomes. Wakahama *et al.* (1983) reported similar phenomenon of heterochromatin accumulation in the genome of drosophila *D. immigrans* group (Wilson *et al.*, 1969, Rangnath and Hagele, 1982). In the present text chromosomal biodiversity has been studied in 9 species of lady bird beetles and will be helpful for base line data for correct identification of Coccinellids.

Heterochromatin has been generally observed in higher organisms, both plants and animals, particularly in the insects of the order Diptera (Heiter, 1928; White, 1978). Constitutive heterochromatin always appears in mitotic chromosomes especially in the pericentric regions, as blocks of dark staining flanking the centromere. In the dipteran insects, constitutive heterochromatin might extend more than 60 per cent of the metaphase length of the X chromosome and as much as 50 per cent of certain autosome pairs (Britten and Kohne 1968, Peacock *et al.*, 1977, 1981, Appels and Peacock 1978, Bonaccorsi and Lohe 1991, Lohe *et al.*, 1993) Further, modern cytogenetic techniques and molecular analysis of constitutive heterochromatin in D. *melanogaster* chromosomes have revealed structural genetic content and its functional role. Recently, some 30 active genes have been found in the heterochromatin region of D. *melanogaster* chromosomes (John and Miklos 1979, Gatti and Pimpinelli 1992). According to Le *et al.* (1995) there are suppressors of forked gene at the proximal region of the X chromosome, the light gene at the distal region of chromosome arm 2L, and the rolled gene within the proximal region of chromosome arm 2R (Berghelle and Dimitri, 1996). However, the Y chromosome contains certain active genes, particularly the fertility factors. Thus, the presence of heterochromatin in eukaryotic chromosomes suggests its significant role in the regulatory function and consorted evolution of genome (Dover 1982, Dover and Flavell 1984, Pardue and Hennig 1990, Irick 1994, Zuckerkandi and Hennig 1995).

The functional role of heterochromatin is the proper recognition and segregation of homologous chromosomes during meiosis. According to John and Miklos (1979) and John (1988) additional copies of repetitive elements of DNA sequences in homologous heterochromatin may be an evolutionary response to additive affects. Heterochromatin differentiation is probably established simultaneously with or even before any other genetic, ecological or behavioral differences that might also contribute to partial reproductive isolation. Selection against the structural heterozygotes in cytological hybrids may be strong enough to initiate a reinforcement process that could lead to species differentiation and speciation (John, 1988).

The dramatic evidences available so far seems to suggest that heterochromatin differentiation often plays an important role in karyotypic evolution in dipteran insects, at least in the oriental region (John, 1988). Although the formulation of models for speciation in relation to heterochromatin differentiation is inevitably in the realm of speculation, the foregoing cytogenetic data indicate some implications of heterochromatin in phylogenetic affinity and consequently the evolutionary diversions of sibling species or closely related species of these dipteran insects. Therefore, detailed investigations in to the dynamics of heterochromatin accumulation, particularly at the molecular level and its evolutionary significance, remains intriguing and challenging in Coccinellids.

Beauchamp and Angus (2006) studied the karyotype of four species of Lady bird beetles (Coleoptera; Coccinellidae). All were clearly different from one another however, two of them *Coccidula* spp. were more similar to one another than to either of the *Rhyzobius* ones. The differences between the *Coccidula* spp. were mainly concerned with the RCLs of the autosomes with the longer autosomes relatively longer and the shorter ones relatively shorter, in *C. rufa* than in *C. scutelata*. In a number of cases the RCLs of equivalent autosomes of the two species were significantly different. The autosomes involved were 1, 2, 3, 6, 7, 8 and 9. Such differences between the karyotypes of two species would only arise as a result of multiple translocation of material between chromosomes. Since karyotype heterozygous for translocations would be unable to pair up for first division of meiosis, individual with such karyotype would be sterile. Therefore, such karyotype differences are very good evidence of the species difference between two *Coccidula*.

Both the species have supposed archetypal polyphagan karyotype of 9 pair of autosomes plus sex chromosomes comprising a normal X and a very small Y, which associates with X via a cytoplasmic vesicle during first meiotic division (Xyp). The situation with the two *Rhyozobius* spp. was very different. The karyotype of *Rh. chrysomeloides* differed from those of the *Coccidula* spp. in having one fewer pair of autosomes. Unlike the two *Coccidula* species, *Rh. chrysomeloides* showed chromosome variation. The karyotype showed heterozygous for a deletion in short arm. This deletion was detectable in intact replicates of the chromosome. According to Sunmer, (2003) this may correspond to the fragile sides found in human and other chromosomes. The chromosomes in two other nuclei from this specimen do not showed the deletion, so it appears to represent spontaneous events in a somatic (Mid-gut) cell.

The karyotypes of two *Rhyzobius* species were totally different from one another. *Rh. chryomeloides* has a fairly conventional karyotype. It differed from those of the *Coccidula* species in having one fewer pair of autosome. If the chromosome formula of the *Coccidula* species (2n = 18 + Xyp) is primitive, then the change to the 2n = 16 + Xyp shown by *Rh. chrysomeloides* is most likely the result of fusion of a pair of autosomes in *Rh. litura,* there is a further reduction in number of autosomes to 16, associated with the development of a Neo-XY system of sex chromosomes.

According to Smith and Virkki (1978) Neo-XY systems are appeared by the fusion of an original X chromosome with an autosome to give neo-X. The unfused replicate of the autosome becomes neo-Y and the XY bivalent at first division of meiosisis held together by chiasmata in the originally autosomal components. Such systems are not uncommon in Adephaga, where the norm is a sex chromosome system involving one X chromosome, females XX, males X0. Development of neo-XY systems is found in beetle species where the norm is Xyp is more complex, either both the X or the Y chromosomes have to fuse with replicates of the same autosome or the Y chromosome must be lost. In some cases these small Y chromosomes were heterochromatic (*Geotrupes mutator* Marsham and *Typhaeus typhoeus* Wilson, Angus, 2004) and thus, carry no genes, but, merely serve as pairing patterns for the X chromosomes. The Coccinellidae, though a polyphagus group with an ancestral Xyp system, have numerous species with neo-XY sex chromosomes (Smith, Virtti, 1978). Nevertheless the discovery of such different karyotype in the two British species of *Rhyzibius* was unexpected. In the present forms Xyp system was not visualized—and worthwhile in future for their utility in biological pest control.

The genus *Rhyzobius* contain 77 species, listed by Korshefsky (1931). Merely all are Austral-asian and only two are European at a first glance at their karyotypes, these two species do not appear to be closely related. Nevertheless, they are the only two species known from Europe. There are one or two listed from the eastern Palaearctic (China), and Smith and Virkki gave the karyotype of *Rh. ventralis*, home Californian, as 2n = 17, eight pairs autosomes plus X0 male. This, at least gave a theoretical starting point for the neo-XY system of *Rh. litura*. All that would be required is fusion of the X with autosomes. The situation regarding the *Rhyzobius* karyotypes may be constructed with that found in the *Coccidula* species where the karyotypes are basically similar with a number of clear species specific differences only. Korshefsky (1931) studied 8 or 9 species (the generic attribution of one species is required), from Eurasia (Palae-arctic Realm) or North America (Nearctic Realm) of the genus *Coccidula* with respect to chromosomal diversity. Coccidula is a small, localized genus, and not old. *Rhyzobius*, on the other hand was either polyphyletic (of multiple origins), perhaps with the Austral-asian species not closely related to the northern Hemisphere ones.

Shaarawi and Angus (1991) made curious comparison of the situation encountered in the karyotypes of the four British *Coccidulinae* and karyotypes of five European species of the genus *Anacaena* Thomson (Coleoptera, Hydrophilidae). Three species were with karyotypes of 2n = 16 + Xyp, one 2n = 14 + Xyp and one has 2n = 10 + Neo-XY. These results are more or less duplicate the finding reported here for a subfamily Coccidulinae, within a single morphologically rather homogeneous genus! Perhaps this should serve as a warning against attributing too much phylogenetic significance to karyotype differences (Shaarawi and Angus (1991).

Epilachna vigintioctopunctata (Fabricius) is phytophagus coccinellid beetle which is serious pest of brinjal, sweet potato and bitter gourd in India. Norio Kobayashi *et al.* (2000) analyzed mitochondrial cytochrome c oxidase I gene sequences (645 bp) for seventeen individuals of *E. vigintioctopunctata* from eight localities in east and southeast Asia revealed that the populations are divided into two genetically distinct groups (Chiba, Tokyo, Naha, Iriomote, Bangkok vs. Kuala Lumpur, Padang, Bogor). The number of nucleotide substitutions between sequences of different groups was 57–60, while that between sequences within each group was 1 – 8 Karyotypes of the two groups were also distinctly different. Crossing experiments showed that there exist strong postmating barriers between the two groups: eggs obtained from between-group crossings usually did not hatch, whereas more than 90 per cent of eggs from within-group crossings hatched. Therefore, they concluded that *E. vigintioctopunctata*, a notorious pest of solanaceous crops in Asia and Australia, is composed of at least two reproductively isolated biological species that probably occupy different geographic ranges.

In predatory Coccinallids, specially *Menochilus sexmaculatus* (Fab) colour variation occurs at larger extent may be due to genetic material. Species are often viewed as reproductively isolated entities (Dobzhansky, 1937; Mayr, 1942). Such "biological species" are not always morphologically distinct. There are many cases of cryptic species that can only be recognized by crossing experiments or molecular characterization (Mayr, 1963; Futuyma, 1997). In phytophagous insects, there has been intense debate over the importance of sympatric speciation (Mayr, 1963; Otto and Endler, 1989; Futuyma, 1997; Howard and Berlocher, 1998).

For this reason, great effort has been expended for characterizing cryptic phytophagous species occurring in geographic areas (sympatric species), whereas less effort has been concentrated on allopatric or parapatric cryptic species. Hence, the literature may exhibit a bias towards sympatric examples. allopatric/parapatric cryptic species. The phytophagous ladybird beetle *E. vigintioctopunctata* is widespread in Asia and Australia, and is allopatric and cryptic species is notorious pests for causing severe damage to solanaceous crops such as eggplants and tomatoes (Dieke, 1947; Li and Cook, 1961; Pang and Mao, 1979; Hoang, 1983; Richards, 1983; Katakura *et al.*, 1988; Richards and Filewood, 1990; Park and Yoon, 1991; Li, 1993; Shirai and Katakura, 1999). Recently, it has been reported on a leguminous weed *Centrosema pubescens* in Java and Sumatra (Nishida *et al.*, 1997; Shirai and Katakura, 1999). This species is known to exhibit a considerable degree of geographic variation in some external features, and thus, it was previously treated as many different species or subspecies (cf. Dieke, 1947). However, most of these variations were in external appearance, *i.e.*, elytral spot patterns and degree of melanism, which are now known to vary even within a single population (Katakura *et al.*, 1988, 1994). Subsequent studies name very close resemblances in important structural characters, including genitalia of both sexes, among the "species" and "subspecies". *E. vigintioctopunctata* is thus currently treated as a widely distributed highly polymorphic species (Li, 1993). In the course of analyzing the phylogenetic relationships of certain species of Asian epilachnines using mitochondrial DNA sequences, Katkura et. al. (1994) found that Japanese and Javanese samples of *E. vigintioctopunctata* have considerable genetic difference comparable to that found between some distinct congeneric species. Preliminary crossing experiments also suggested the presence of a strong post-mating barrier between the two populations.

The chromosome composition of *E. vigintioctopunctata* was 2n=18 with male heterogametic XYp-XX in sex determination. It was confirmed by Kobayashi et. al. (2000) as the basic chromosome number of *E. vigintioctopunctata* although some individuals from Padang and Bogor possessed supernumerary chromosomes (2n=19, 20, 21, 22). The two groups of samples recognized on the basis of the mtDNA analyzed by them were also differed from each other karyologically. The specimens from Japanese localities and Bangkok possessed an obviously larger Y chromosome compared with those from Kuala Lumpur and Indonesian localities ("southern populations"). On the other hand, there was no detectable difference in the size of the X chromosome between the two groups. Owing to these facts, two distinctly different types of the X-Y sex bivalents were observed in the first meiotic metaphases: XYp, exclusively found in the "northern populations" and Xyp, in the "southern populations". The karyotypes of the two groups were strikingly different from one another in the arm ratio of each chromosome of the complement as well. Namely, smaller six autosomes were acro- or subtelocentric in the "northern populations" and meta type of chromosome configuration was typical for the diphasic chromosomes often encountered in various groups of Coleoptera including Epilachna (Drets *et al.*, 1983; Tsurusaki *et al.*, 1993), and the block can be inferred to be heterochromatic, although no C-banding technique was employed. The mode of karyotypic differentiation between the two forms is highly reminiscent of that found between two closely related forms of the *E. vigintioctomaculata* species complex (Tsurusaki *et al.*, 1993).

E. *vigintioctopunctata* is currently treated as a widely distributed highly polymorphic species (Li, 1993). In *E. vigintioctopunctata*, there exist large inter populational variations in spot patterns and some other external features (Abbas et. al., 1988; Katakura *et al.*, 1988, 1994), but all of them share identically shaped genitalia, a characteristic of *E. vigintioctopunctata* (Katakura *et al.*, 1988). The samples of *E. vigintioctopunctata* studied were composed of two biological species. One of them, here tentatively called "the northern form", was consisted of the samples from Japan through Thailand (Bangkok), and the other one, "the southern form", was comprised of samples from the peninsular part of Malaysia to the two large islands (Sumatra, Java) of Indonesia. The two forms were genetically and karyologically well differentiated. Furthermore, they were potentially reproductively isolated from each other by a very strong postmating barrier. They suggested that there might be a certain degree of sexual isolation, since some femals did not retain sperm when they were kept for nearly four weeks with males of the other form. Available evidence suggests that the two forms were allopatric or parapatric, and the two forms seem to replace to each other in the Malay Peninsula, somewhere between Bangkok and Kuala Lumpur. If this interpretation is correct *E. vigintioctopunctata*, a common pest species widespread in East and Southeast Asia, may provide an extremely suitable situation for the studies of various controversial issues of evolutionary biology, in particular those concerning the mode of speciation and reinforcement of reproductive isolation. Such studies are waiting in the Coccinellidae family of order Coleoptera in India.

Amanda *et al.* (2009) analyzed mitotic and meiotic chromosomes of several populations of *Eurysternus caribaeus* (Coleoptera; Scarabaeidae) through conventional staining, C-banding, base-fluorochromes, silvernitrate staining and fluorescent *in situ* hybridization (FISH). All specimens showed 2n = 8 in their karyotypes with a Neo-XY sex system (Y is a submetacentric and X a metacentric) and three pair of submetacentric autosomes. The analysis of constitutitive heterochromatin (CH) revealed small blocks located in the centromeric region of all chromosomes which do not present positive staining under the flurochromes CMA3 and DAPI. Silver nitrate staining revealed that the nucleaolar organizer region (NORs) was associated with the sex chromosomes. The FISH technique revealed that rDNA cited in the X and Y were different in size. Data from different populations indicates that the diploid number reduction (2n = 8) observed in *E. caribaeus* was established. More over, this reduction occasioned the translocation of rDNA sites to the sex chromosomes, X and Y. An uncommon pattern in Scarabaeidae that was observed for the first time by the FISH technique (Amanda *et al.,* 2009).

The family Scarabaeidae is cosmopolitan which comprises approximately 2000 genera and 25,000 species. In Neotropical region about 4706 species have been recorded and 1777 species were identified in Brazil (Costa, 2000). Some of their representatives showed important functions as pollinators of plants, organic matter recyclers and biological controllers of agricultural pests and as indicators for the analysis of biodiversity in tropical forests (Halffter and favila 1993). Despite the large number of species, chromosomal diversity of Scarabaeidae was studied only in 390 species, 1.57 per cent of species have been analyzed by using conventional staining. The family showed conserved karyotype with more than 50 per cent of the species presenting the diploid number 2n = 20, Xyp sex mechanism and biarmed chromosomes. This condition has been considered primitive to this group and also to the whole order Coleoptera (Smith and

Virkki, 1978, Yadav *et al.*, 1979, Martins 1994, Moura *et al.*, 2003, Bione *et al.*, 2005a,b, Cabral-de-Mello *et al.*, 2008). *E. caribaeus* has the lowest (2n = 8), while *Autoserica assamensis* the highest diploid number for the family (2n = 30) (Yadav *et al.*, 1979, Cabral-de-Mello *et al.*, 2007). The family presents seven sex mechanisms (XY, Xy, XYp, Xyp, Xyr, X0 and Neo-XY) with variations in the chromosomal morphology in some species (Yadav *et al.*, 1979, Moura *et al.*, 2003, Bione *et al.*, 2005a,b, Cabral-de-Mello *et al.*, 2007, Dutrillaux *et al.*, 2007). Such studies are waiting in the Coccinellidae family of order Coleoptera.

Karyotypic structural variation has been reported in some Coleopteran families, Buprestidae (2n = 12 to 2n = 46), Elateridae (2n = 4 to 2n = 23), Lamprydae (2n = 19, X0) and Cantharidae (2n = 13, X0) showed specific number of chromosomes (Smith and Virkki 1978, Machado *et al.*, 2001, Karagyan *et al.*, 2004, Dias *et al.*, 2007, Schneider *et al.* (2007a). In the present forms chromosomal number was up to 12 to 20 in various species. However, in the insect class, as a whole, there are groups that present modal and conserved diploid number *i.e.* acridid grasshopper (Orthoptera), Libellulliae (Odonata) and groups less covered showing variation in number of autosomes and/or sex chromosomes in Phaneropterinae (Tettigoniidae, Orhoptera) and Reduviidae (Heteroptera) (Hewitt 1979, Mola *et al.*, 1999, Ferreira and Mesa 2007, Poggio *et al.*, 2007).

Angus *et al.* (2007) studied about 70 Scarabaeidae species using differential or molecular cytogenetic techniques, such as C-banding, base specific fluorochromes, Silver nitrate staining or fluorescence *in situ* hybridization (FISH) (Moura *et al.*, 2003, Wilson and Angus 2004, 2005, 2006, Bione *et al.*, 2005a,b, Dutrillaux *et al.*, 2007). The constitutive heterochromatin (CH) in this family was predominately located in the pericentric region of the chromosome and this genomic component showed wide heterogeneity regarding AT-richness and GCrichness moreover, species with telomeric, interstitial CH and diphasic chromosomes have been described in this family (Columba *et al.*, 1996, 2000, 2006, Moura *et al.*, 2003, Bione *et al.*, 2005a, Angus *et al.*, 2007, Macaisne *et al.*, 2006). In the present forms chromosomes were of 7 types. The nucleolar organizer regions (NORs) are predominately located in a single autosomal pair or in X chromosome. However, some species showed more than one rDNA sites clustered in different chromosome pairs (Moura *et al.*, 2003, Bione *et al.*, 2005, Macaisne *et al.*, 2006).

Nutan Karnik *et al.* (2010) studied karyotype instability in the Ponerine ant genus *Diacamma*. Ants are genrally classified as highly eusocial species in which the queen and worker castes are morphologically differentiated (Wilson 1971, Holldobler and Wilson, 1990). However, about 100 species belonging to the phylogenetically and morphologically primitive subfamily Ponerinae which lack a morphologically distinguishable queen caste (Wheeler 1915, Peeters 1991). In these species, workers have retained the ability to mate and reproduce. In the queenless ponerine ant genus *Diacamma*, reproductive monopoly is achieved by a unique mechanism. Here, all individuals are morphologically identical and enclose with a pair of rudimentary, mesothorasic wing buds called gemmae which apparently release an exocrine signal (Tulloch 1934, Peeters and Billen 1991, Baratte *et al.*, 2006a). Gemmae, anable individuals to perform sexual calling and are thus necessary for mating to occur. The Gamergate (mated egg laying worker), (Peeters, 1993) however multilets the gemmae of all the eclosing individual who eclose after her (Fukumoto *et al.*, 1989, Peeters and Higashi 1989). According to Baratte *et al.*, 2006b mutilation of the gemmae leads to irreversible nurological changes in the workers and they lose their ability to perform sexual calling and thus they can not mate (Gronenberg and Peters 1993). They further say that multilated workers do not multilate others, so that after the death of gamergate, the first worker to eclose retains her gemmae and assumes the role of the gamergate. There is also an interesting evidence that cues for multilation originate in the callows, presumably in the gemmae themselves (Ramaswamy *et al.*, 2004). Surprisingly, in some *Diacamma* populations from South India, tentatively called *Diacamma* sp. from Nilgiri (Hereafter referred to as 'nilgiri'), the gamergate does not multilate her nest, mates and yet monopolizes reproduction by using dominance interactions (Peeters *et al.*, 1992) in which 'nilgiri' is mislabeled as *Diacamma vagans*).

Baudry *et al.* (2003) studied the molecular phylogeny of the *Diacamma* genus that *Diacamma ceylonensei* is closely related to 'nilgiri') and indicates that 'nilgiri' originates from the most recent diversions in the tree (Veuille *et al.*, 1999). These taxa are almost similar in morphology but for the mutilation of gemmae it is in *D. ceylonensei* and not in 'nilgiri'. In addition to the behavioral difference related to the mutilation of gemmae, microsatellite and mitochondrial markers have revealed significant genetic diversions between these taxa (Baudry *et al.*, 2003). For all the above mentioned reasons *D. ceylonense* and 'nilgiri' provides an interesting model system for the study of incipient speciation. King (1993) says that karyotype is generally an invariant character of each species and is therefore considered to be a taxonomic value. Nevertheless, chromosomal rearrangements often accompany events of speciation in order to produce species-specific karyotypes. It follows that the study of karyotypic differences among taxonomically closely related species could provide insights in to the mechanism of speciation (John 1981). Within the family coccinellidae two groups are visualized, one phytophagus, coccinellinius and another is coccinellids which is predaceous. The chromosomal studies will be helful for correct identification of the species since both groups are strikingly similar to each other by morphological features.

Correa *et al.* (2008) studied *Zabrotes subfasciatus* (Boh.) in its agronomic and biochemical aspects due to its importance as a damaging insect to leguminous grains during storage. The few cytogenetic studies published on this species yielded conflicting results. Hence, karyotype was analyzed in order to accurately describe the chromosome C-banding patterns and meiosis. The brain ganglion at the prepupa and the adult and pupal testes were analyzed. All individuals had 26 chromosomes in both brain ganglion and spermatogonic mitotic metaphases. These chromosomes were classified as follows: the 12th pair and the Y chromosome were telocentric; the X chromosome was acrocentric; the 4th and 5th pairs were submetacentric and the remaining pairs were all metacentric. One of the members of the 5th pair presented a secondary constriction. All chromosomes presented pericentromeric heterochromatin. The large arms of the pairs, 9 and X presented heterochromatin. The X chromosome heteropyknotic throughout the prophase of the first meiotic division. The subphases of prophase I were a typical and meiosis II was rarely identified. Testes of all males showed a few cells; the bivalents were rod-like shaped in metaphase I. Karyological formulae were 2n = 24 + XX in females and 2n = 24 + XYp and either n = 12 + X or n = 12 + Y in males.

The cerebral ganglia of 15 male and 19 female pupae were analyzed, enabling the observation of 21 metaphases in males and 28 in females. All individuals analyzed from both populations presented a diploid number of 26 chromosomes. The Morphometric analysis showed chromosomes of three distinct sizes according to the total relative length (T): large, T > 10; medium, < T < 10; and small, T <. The autosomes were classified as telocentric (12th pair), submetacentric (4th and 5th pairs) and metacentric (the remaining pairs). One chromosome of the 5th pair showed a secondary constriction possibly corresponding to one NOR. This constriction was evident for both homologous chromosomes in females' metaphases. In the females, the sexual pair is formed by homologous chromosomes while males present a large X chromosome and a small Y chromosome. The sex chromosomes were classified as acrocentric (X) and telocentric (y). The karyotype of Z. *subfasciatus* was symmetrical, because the 12th pair and the allosomes showed a discrepant size in relation to the other eleven chromosome pairs that presented a gradual size variation. Smith and Virkki (1978) suggested that the symmetrical karyotypes appeared secondarily in Coleoptera. Further cytogenetic studies including other bruchids as well as a resolved phylogeny for the family will enhance our ability to elucidate whether Z. *subfasciatus* is a karyotypically conserved species. Z. *subfasciatus* was first cytogenetically described by Minouchi (1935), who used tissue slices impregnated with paraffin and presented the male karyological formula 2n = 12II + XO + 0/1 extra chromosome. That description was confirmed by Smith and Virkki (1978). However, Takenouchi (1972) using the smearing technique did not find any B chromosomes and observed the karyotypes 2n = 24 + XX in females and 2n = 24 + Xyp in males.

There was karyotypic variation within Z. *subfasciatus* that the only chromosome which was not entirely homologous behaved cyclically and presented a "parachute" structure in males that indicates a Xyp-type sex determination in this species. The most characteristic male sex-determining system of Coleoptera, the "parachute-like" Xyp, represents a non-chiasmatic association of a generally metacentric X and a small and mostly metacentric y-chromosome (Petitpierre *et al.*, 2004; Palomeque *et al.*, 2005). Rozec (1994) and Lachowska and Holecova (2000), analyzed the chromosomes in three beetle species in the genus *Phyllobius*, they found the Xyp system of sex determination and as the most common form in Coleoptera. This "parachute" conformation of sexual chromosomes was associated with the presence of the nucleolar organizing region sites (NORs)

(John and Lewis 1960, Smith and Virki 1978). Maffei *et al.* (2001) mapped rDNA genes in meiotic metaphases of *Olla v-nigrum*, using FISH and Ag-NOR staining and identified active NORs in the sex bivalent during meiosis. The "parachute" configuration of sexual chromosomes has been reported for other coleopterans such as *Eriopis connexa* Mulsant (Mafei *et al.*, 2000); some species in the genus *Chrysolina* (Chrysomelidae) (Petitpierre *et al.*, 2004, Palomeque *et al.*, 2005); several species in the genus *Timarcha* (Chrysomelidae) (Gomez- Zurita *et al.*, 2004); in *Phyllophaga* (*Phytalus*) *vestita* (Moser) and *Phyllophaga* sp. *aff. Capillata* (Scarabeidae) (Moura *et al.*, 2003); various species in the genus *Cyrtonus* (Chrysomelidae) (Petitpierre and Garneria 2003); in *Epicauta atomaria* (Germar) (Meloidae) and *Palembus dermestoides* (Fairm.) (Tenebrionidae) (Almeida *et al.*, 2000); and also for some species in the genus *Monochamus* (Cerambicydae) (Cesari *et al.*, 2004). However, there has been some questioning of the nucleolar theory by other investigators based on evidence of the occurrence of nucleoli in a pair of autosomes in other species of Coleoptera (Maffei *et al.*, 2004).

Forer (1980) verified that in crane flies the autosomes separate first than the sex chromosomes, but at the end of the meiosis, the sexual chromosomes segregate normally, each going to a respective cell. In the case of *Z. subfasciatus*, was time-lagged phenomenon occured in the migration of the sexual chromosomes. However, the separation should proceed normally as in metaphase II, cells with only one sexual chromosome were always observed. The subphases of prophase I were somewhat distinct while, the phases of meiosis II were difficult to identify. Testes tissue presented just a few meiotic cells as described by Takenouchi (1972). In diplotene and diakinesis, the chromosomes occurred as a mass. According to Smith and Virkki (1978), the Coleoptera are not favorable for crossing over studies because the diplotene was commonly diffused, which made difficult to observe the chiasma terminalization process. In metaphase I, the chromosomes were connected in the extremities. Meiotic cells were not found in the ovaries analyzed. Observations of two female ovaries indicated that the oögonium metaphase has invariably 26 chromosomes. All individuals exhibited the karyotypes 2n = 24 + XX in females as well as 2n = 24 + Xyp, and either n = 12 + X or n = 12 + y in males.

Chromosomal studies of Buprestid beetles are widely attempted in the world (Smith and Virkki 1978; Karagyan and Kuznetsova 2000; Karagyan 2001; Karagyan *et al.*, 2004; Karagyan and Lachowska 2007; Moura *et al.*, 2008). These beetles belong to the family Buprestidae (Coleoptera) is one of the largest groups of Polyphagan beetles containing over than 14500 nominal species worldwide (Bellamy 2008a-d, 2009). Until now, karyotypes have been published for 88 species (34 from Armenia) of jewel- beetles belonging to 22 genera and 14 tribes of the subfamilies Julodinae, Polycestinae, Chrysochroinae, Buprestinae and Agrilinae. In the works listed the data were obtained by study of gonads of imagos. Only one work was based on the study of hemocytes of the larva of *Chalcophora mariana* (Linnaeus, 1758) (Chrysochroinae, Chrysochroini) with diploid chromosome number 2n=22 (Baragano, 1978). The diploid chromosome numbers (2n) in the family Buprestidae ranged between 12 and 46. The modal number was 2n=20 found in 16 species, 8 genera, 6 tribes and 4 subfamilies. The most frequent sex chromosome system was XX/XY. The XY system of males was diverse (Xyp, Xyr, "XY", neo-XY and multiple X- and Y- sex chromosomes). The Xyp "parachute" type was the most common and occured in 64 species, 15 genera, 10 tribes and 4 subfamilies.

In *Acmaeodera pilosellae persica*, the chromosome number was 2n=20, n= 9 + neo-XY. In prometaphase I/metaphase I nine autosomal bivalents formed a series of gradually decreasing sizes and large heteromorphic neo-XY sex-bivalent In late diakinesis/prometaphase I the 5–6 rod-shaped bivalents have most likely a terminal chiasma and 3–4 ring-shaped bivalents – twochiasmata. The X-chromosome was submetacentric, Y-chromosome was most probably acrocentric and similar in size to the shorter arm of X-chromosome.

The karyotype of *Sph. scovitzii* studied using Ag-banding (Karagyan 2001) and the data on karyotype of this species obtained in that study were confirmed. Unfortunately, until now, the male diploid karyotype of the species can not be determined with certainty. Thus, it seems that karyotype consists of 38–46 chromosomes, most probably of 46; the sex chromosomes could not be identified. Some of chromosomes have a very small amount of constitutive heterochromatin weakly visible in pericentromeric regions and do not form distinct blocks. The DAPI staining of chromosomes did not reveal any positive signal, yet fluorescence after CMA3 staining was discovered. In metaphase I, three or sometimes four rodshaped bivalents showed distinct CMA3 positive signals on telomeric regions of both homologues. These signals were quite stable in the largest and in two of the middle-sized bivalents and correspond to Ag-positive material revealed by the AgNOR banding technique. Weak CMA3 positive signals were also at times visible in some other small bivalents.

Male *Dicerca aenea validiuscula* displayed 20 chromosomes including 9 autosomal pairs and X- and Y- sex chromosomes in mitotic metaphase. All autosomes were biarmed: one pair large and 5 pairs of middle-sized metacentrics, 2 pairs of middle-sized submetacentrics and 1 pair of middle-sized subtelocentric. The X-chromosome was middle-sized and acrocentric, Y-chromosome was dotlike with unclear morphology. In mitotic metaphase CMA3 positive signals were found in the pericentromeric region of long arm of middle-sized homologous pair of subtelocentric autosomes. The DAPI staining did not reveal any positive signal. In diakinesis/prometaphase I nine autosomal bivalents and heteromorphic sexbivalent most probably of "parachute" Xyp type were observed. The autosomal bivalents formed a series of gradually decreasing sizes. There were 5–6 ringshaped autosomal bivalents with two chiasmata, 1–2 rod-shaped bivalents possessed most likely one terminal chiasma and 1–2 were cross-shaped with an interstitial chiasma. The heterovalent Xyp was rather small.

In *Sphaerobothris aghababiani*, diakinesis/prometaphase I seven autosomal bivalents and a sex chromosome heterovalent of the Xyp type were observed. The bivalents gradually decreased in size. The Xyp sex heterovalent was smallest element in the set. The majority of autosomal bivalents were rod-shaped, but in some cells one or two ring-shaped autosomal bivalents were observed. In total, the karyotypes of 92 species of jewel-beetles belonging to 23 genera, 14 tribes of 5 subfamilies have been studied (Smith and Virkki 1978; James and Angus 2007; Schneider *et al.*, 2007; Cabral-de-Mello *et al.*, 2008; *et al.*). Generalization of data including new ones showed that the modal diploid chromosome number in Buprestidae was 2n=20 (9AA + X- and Y- sex chromosome heterovalent in males) so far found in 18 species belonging to 8 genera, 6 tribes, 4 subfamilies. The Xyp type of sex chromosome heterovalent was modal and occured in 66 species, 16 genera, 10 tribes, 4 subfamilies. The most common karyotype 2n=20 (19+Xyp) was considered as modal. The new data confirm modality of this karyotype within the family. This karyotype occured in a large number of beetles from different families.

Even this restricted material showed noticeable variability of distribution of argentophilic material (probably NOR) in the karyotypes of jewel-beetles. The argentophilic material was located on:

1. The autosomes of *Sphenoptera scovitzii* (2n=38–46) and *Sph. mesopotamica* Marseul, 1865 (2n=24, Xyp),

2. Both on the sex chromosomes and on the autosomes of *Acmaeoderella villosula* Steven, 1830 *Acmaeoderella boryi* Brullé, 1832 in Karagyan 2001 (2n=18, Xyr), *Acmaeodera pilosellae persica* (2n=20, neo-XY), *Sphaerobothris aghababiani* (2n=16, Xyp) and *Dicerca aenea validiuscula* (2n=20, Xyp), and

3. Sex chromosomes of bivalent of *Acmaeoderella flavofasciata* (Piller and Mitterpacher, 1783) (2n=18, Xyr), *A. gibbulosa* Menetries, 1832 (2n=18, Xyr), *A. vetusta* (Menetries, 1832) (2n=18, Xy). In *Euchroma gigantea* 2n=32, 36, XXXYYY – (Moura *et al.*, 2008) and Sex chromosomes only, situating either on one of the sex chromosomes or localized between argentophilic material labeled the multiple sex chromosomes chain.

In Coleoptera NORs may be located in some autosomal pair and/or sex chromosomes (Almeida *et al.*, 2000; Moura *et al.*, 2003; Bione *et al.*, 2005b; Holecová *et al.*, 2008). The most common pattern in Coleoptera was the location of the nucleolus organizer region in one autosomal pair (Virkki 1983; Virkki *et al.*, 1984; Vitturi *et al.*, 1999; Colomba *et al.*, 2000; Moura *et al.*, 2003; Almeida *et al.*, 2006; Schneider 2007; *et al.*). On the other hand, the argentophilous body has been repeatedly observed between the sex-chromosomes of Xyptype (Postiglioni and Brum-Zorrilla, 1988; Virkki *et al.*, 1990, 1991; Maffei *et al.*, 2001; Moura *et al.*, 2003 *et al.*). Among Buprestidae, the whole Xyp sex bivalent was brightly argentophilic and beared NOR only in *Sphaerobothris aghababiani* and *Dicerca aenea validiuscula*. The lack of relationship between nucleolus and sex chromosome system of Xyp has also been demonstrated for some

Coleoptera including one species of jewel-beetles, *Sphenoptera mesopotamica* (Karagyan 2001). According to some authors, the "parachute" can be strongly marked by the silver nitrate during different phases of meiosis independent of whether or not the NORs are located in Xyp bivalent (Virkki *et al.*, 1991; Moura *et al.*, 2003; Bione *et al.*, 2005b; Mendes-Neto *et al.*, 2010). This phenomenon was probably related to the presence of argentophilic substance (proteins) that theoretically facilitates the configuration and segregation of the sex chromosomes of the Xyp system (Virkki *et al.*, 1990, 1991; Juan *et al.*, 1993; Petitpierre, 1996; Moura *et al.*, 2003; Bione *et al.*, 2005a, b; Schneider *et al.*, 2007).

Chromosome staining by DNA base specific fluorochromes has been used in cytogenetic studies of Coleoptera (Juan and Petitpierre, 1989; Vitturi *et al.*, 1999; Colomba *et al.*, 2000; Moura *et al.*, 2003; Schneider *et al.*, 2007; Lachowska 2008; Mendes-Neto *et al.*, 2010) including Buprestidae. The correlation between NORs and CMA3 bands was quite common in some insects, including beetles (Camacho *et al.*, 1991; Vitturi *et al.*, 1999; Colomba *et al.*, 2000; Maffei *et al.*, 2001; Kuznetsova *et al.*, 2001; Araújo *et al.*, 2002; Brito *et al.*, 2003; Nechayeva *et al.*, 2004). Silver staining mainly revealed transcriptionally active NORs (Sumner 1990), as opposed to fluorochrome CMA3 staining which labels NORs independently of their activity (Schmid and Guttenbach, 1988). In this study, the fluorescent signals after CMA3 staining was positive in all four species of jewel beetles. While in *Sphenoptera scovitzii* and *Sphaerobothris aghababiani* they were nearly fully correlated with argentophilic material observed on silver dyed chromosomes. In *Acmaeodera pilosellae persica* CMA3 signals correlated with argentophilic blocked on 2–3 autosomal bivalents but not with Y-chromosome which was dyed argentophilic as well. Meanwhile, in *Dicerca aenea validiuscula* CMA3 signal was correlated with one of autosomal pairs when argentophilic material was revealed in the same autosomal pair as well as on sex bivalent. In conclusion, seen data offer important insights into the karyotypes characteristics of jewel-beetles which may be useful in elucidation of relationships both among the species of the family itself as well as between jewel-beetles and the representatives of other coleopteran families.

Robert (2012) studied the current usage of the name *Aphodius fimetarius* (Linnaeus, 1758) for a Holarctic species of aphodiine dung beetle. Since a different species has been erroneously designated as the lectotype, it was proposed that the previous type fixation for the species *Aphodius fimetarius* (Linnaeus, 1758) was set aside and a neotype consistent with the current usage was designated. The species diagnostic morphological characters showed variation overlapping with those of the most similar species, *Aphodius pedellus.* Robert (2012) suggested a modern, chromosomally determined specimen as the neotype. According to Wilson (2001) *Aphodius fimetarius* (Linnaeus, 1758), was used by all authors in the preceding hundred years, comprised two species clearly separable on their karyotypes and reported on relevant type material and designated lectotypes for *Scarabaeus fimetarius* Linnaeus, 1758 and *Scarabaeus pedellus* De Geer, 1774.

Petitpierre (2011) analyzed chromosomally nearly 260 taxa and chromosomal races of subfamily Chrysomelinae which showed a wide range of diploid numbers from 2n = 12 to 2n = 50, and four types of male sex-chromosome systems with the parachute-like ones Xyp and XYp clearly prevailing (79.0 per cent), but with the XO well represented too (19.75 per cent). The modal haploid number for chrysomelines was n = 12 (34.2 per cent) although it was not presumed most plesiomorph for the whole subfamily, because in tribe Timarchini the modal number was n = 10 (53.6 per cent) and in subtribe Chrysomelina n = 17 (65.7 per cent). Some well sampled genera, such as *Timarcha, Chrysolina* and *Cyrtonus*, were variable in diploid numbers, whereas others, like *Chrysomela, Paropsisterna, Oreina* and *Leptinotarsa*, were conservative. The main shifts in the chromosomal evolution of Chrysomelinae seems to be centric fissions and pericentric inversions but other changes as centric fusions are also clearly demonstrated. The biarmed chromosome shape was the prevalent condition, as found in most Coleoptera, although a fair number of species hold a few uniarmed chromosomes at least. A significant negative correlation between the haploid numbers and the asymmetry in size of karyotypes (r = -0.74) has been found from a large sample of 63 checked species of ten different genera. Therefore, the increases in haploid number were generally associated with a higher karyotype symmetry.

The sub family Chrysomelinae showed a wide variation of diploid chromosome numbers and meioformulas, from 2n = 12 and 5 + neo XY, respectively, in the South American *Doryphora quadrisignata* (Vidal, 1984), to 2n = 50 and 24 + Xyp in the European *Chrysolina rufoaenea* (Petitpierre and Mikhailov, 2009). These shifted number were almost always due to structural chromosome rearrangements, because only a few polyploidy parthenotes have been recognized to date, all of them restricted to the genus *Calligrapha*, (Robertson, 1966; Smith and Virkki, 1978). The range of variation of haploid numbers for the total 259 taxa and chromosomal races in the 38 examined genera, showed an almost continuous list of numbers but with a modal one of n = 12 (34.2 per cent), followed by three others of n = 10 and n = 17 (both with 9.6 per cent), and n = 20 (7.6 per cent). Conversely, the parachute-like sex-chromosome system (Xyp) of a nonchiasmate nature was clearly prevailing in the subfamily (79.0 per cent). This system consisted mostly of a large X and a small y-chromosome, looking such as this configuration at metaphase I, or more rarely, two large X and Y chromosomes (XYp), held together by a non-nucleolar argyrophilic substance (Virkki 1984; Postiglioni and Brum-Zorrilla, 1988; Virkki *et al.*, 1991). The Xyp was probably the most plesiomorphous condition in Chrysomelinae, as it was for the whole beetles of the suborder Polyphaga (Smith, 1951; Smith, 1952; Smith and Virkki, 1978) while, the others so far checked in the subfamily, the XO (19.75 per cent) and neoXY or XY systems (1.2 per cent), are certainly derived.

The modal number of n = 12 chromosomes has been found in five out of the six reported subtribes, it was very seldom in Timarchina and Chrysomelina, and it does not occur to date in the poorly surveyed Entomoscelina, with only seven analyzed species (Petitpierre and Grobbelaar, 2004; Petitpierre unpublished), belonging to five among the 27 described genera (Daccordi, 1994).

The Timarchina subtribe showed a striking modal value of n = 10, and 9 + Xyp meioformula, which were the modal and presumably the possible plesiomorphous state for this group, as well as for the whole beetles of the suborder Polyphaga (Smith, 1952; Smith and Virkki, 1978; Angus *et al.*, 2007). The two most ancestral extant subgenera of *Timarcha*, *Americanotimarcha* and *Metallotimarcha*, both on morphological and molecular grounds (Iablokoff-Khnzorian, 1966; Jolivet, 1989; Gómez-Zurita *et al.*, 2000; Gómez-Zurita, 2004), comprised only species showing the highest diploid numbers found in the genus, 2n = 38 and 2n = 44 (Petitpierre and Jolivet, 1976; Jolivet and Petitpierre, 1992). If these high numbers were the possible plesiomorphous condition for the chromosomal evolution in *Timarcha*, how could have derived all the common 20-chromosome species by independent processes? The most parsimonious view would be assuming a hypothetic stem species for the genus, represented with a karyotype of 20-chromosomes, from which the ancient ancestors of the three extant subgenera would have splitted. The *Americanotimarcha* and *Metallotimarcha* through multiple chromosome fissions, followed by pericentric inversions and/or chromatin accretions of uniarmed elements, to recover some of them later to their ancient biarmed condition, while within the species-rich *Timarcha* s.str. subgenus much more conservative events of chromosomal shifts had presumably occurred in the karyological origin of most species.

The subtribe Doryphorina displayed a 2n(male) = 35 modal chromosome number and 17 + XO meioformula, but this was attributed to a biased sampling on the species of *Leptinotarsa*, all but one sharing these values (Hsiao and Hsiao 1983). However, the species of eight genera of analyzed Doryphorina, out of the two closely related in the genus *Labidomera*, have karyotypes of much lower chromosome numbers, namely, n = 12 in six species of five different genera, *Desmogramma*, *Leucocera*, *Strichosa*, *Platyphora* and *Zygogramma*, a fact which could possibly hint towards the supposed most plesiomorphous karyotype condition for this subtribe. On the contrary, in subtribe Chrysomelina the modal number and meioformula were 2n = 34 and 16 + Xyp, respectively, shared by 65.7 per cent of the 35 surveyed species in twelve genera, and should possibly the ancestral condition (Petitpierre and Segarra, 1985) for this taxon, but it was more uncertain due to the absence of this 2n = 34(Xyp) karyotype and meioformula in half of the twelve sampled genera showed high chromosomal diversity, as measured by standard deviation (SD) of their male diploid chromosome numbers, for example *Chrysolina* with SD = 8.67 in 72 sampled taxa and chromosomal races, *Timarcha* with SD = 4.33 in 42 taxa, and *Cyrtonus* with SD = 6.33 in 15 taxa, whereas other genera have zero or a low diversity such as *Paropsisterna* with SD = 0 in 10 taxa, *Chrysomela* with SD = 0 in 9 taxa, *Oreina* with SD = 1.15 in 12 taxa, and *Leptinotarsa* with SD = 2.77 in 16 taxa. The differences between "variable" and "conservative" genera in their chromosome numbers, were tentatively explained according with the ability for dispersal of flying vs.

flightless species genera, and the number of host-plant families they are able to feed, being both characters in a presumed relationship with the size of local populations and thereby with the chances of fixation for new chromosomal shifts (Petitpierre *et al.*, 1993). The chromosomes may show a huge variable morphology in size and shape, some species have karyotypes made of very few chromosomes of a large size while others have karyotypes of many small chromosomes and there are not evidences of any advantages of ones over others, although minute chromosomes are more easily lost at meiosis if a chiasma fails to be formed, and very large acroor telocentric chromosomes can be cut across before they have been properly separated at anaphase (Sumner, 2003). Chromosomes are elements of the genetic system that may supply worth features to explain evolution among closely related species (White, 1973; King, 1993).

Some 80 among the 259 presently know taxa or chromosomal races of chrysomelines have been examined at the level of ß-karyology *i.e.* by identifying size and shape of individual chromosomes in each karyotype. Such kind of studies have been mainly carried out in certain genera, the North American *Calligrapha* (Robertson 1966) and *Leptinotarsa* (Hsiao and Hsiao 1983), and the Holarctic *Timarcha* (Petitpierre 1970, 1976; Gómez-Zurita *et al.*, 2004), and the Palaearctic *Chrysolina* (Petitpierre 1981, 1983, 1999a, 1999b; Petitpierre *et al.*, 2004; Petitpierre and Mikhailov 2009) and *Cyrtonus* (Petitpierre and Garneria 2003). The karyotypes of chrysomelines were usually composed of meta- or submetacentric chromosomes as occured mostly in all groups of Coleoptera (Smith and, Virkki 1978; Virkki, 1984). This means that the shifting in number due to centric fissions, should necessarily rebuild the emerging acrocentric chromosomes into biarmed ones by pericentric inversions or heterochromatin accretions (Virkki, 1984; Virkki and Santiago-Blay 1993), and this secondary metacentry has been described in diphasic chromosomes of several beetle species (Virkki, 1984). Taking into account the biarmed shape of most chromosomes in chrysomelines, and in other beetles in general, it was evident that the number of major chromosome arms (FN = fundamental number) could not remain constant and increase accordingly with the diploid number. However, many species of high diploid numbers have at least a few acrocentric or subacrocentric chromosomes, which may be the ancient survivors of primary shifts by centric fissions. For instance, the Nearctic

Timarcha intricata with 2n = 44 has 15 of such autosome pair (Petitpierre and Jolivet, 1976; Petitpierre, 1988), *Leptinotarsa lineolata*, *Leptinotarsa behrensi* and *Leptinotarsa decemlineata* (the potato beetle), all with 2n (male) = 35, have seven, four and three, respectively (Hsiao and Hsiao 1983), the Palaearctics *Timarcha pimelioides* with 2n = 28 has five (Petitpierre, 1976, 1988), *Chrysolina gypsophilae* with 2n = 32 has three (Petitpierre, 1999b), *Chrysolina diluta* with 2n = 36 and *Chrysolina haemoptera* with 2n = 40 have four (Petitpierre, 1988), *Chrysolina lepida* with 2n = 42 has six, whereas its closely related *Chrysolina fuliginosa*, also with 2n = 42, has none (Petitpierre 1999a). The extreme cases were those of the European *Chrysolina carnifex* and *Chrysolina interstincta* both with 2n = 40 and having only acrocentric chromosomes, contrary to *Chrysolina helopioides* with 2n (male) = 47 and lacking any of them (Petitpierre, 1981; Petitpierre, 1999a; Petitpierre and Segarra, 1985; Petitpierre *et al.*, 2004). In conclusion, the FN even in species having similar numbers as the latter, can be strikingly distinct, FN = 40 in *Chrysolina carnifex* and *Chrysolina interstincta*, and FN = 94 in *Chrysolina helopioides*. Additional examples of frequent increases of acrocentric chromosomes in Polyphaga beetles associated with high diploid numbers were also reported in Buprestidae (Karagyan and Lachowska, 2007) and in Curculionidae (Lachowska *et al.*, 1998).

Karyotypes can also be classified as symmetrical in size when all chromosomes have similar magnitudes, and asymmetrical when there are two clearly distinct size classes, and these two alternatives can also be applied to chromosome shape, uniarmed chromosomes for asymmetrical and biarmed ones for symmetrical karyotypes (Stebbins, 1971; White 1973). For the sake of simplicity one can consider the asymmetry vs. symmetry in chromosome size but not in shape. The karyotypes of Chrysomelinae offer examples of both types but more often of intermediate states, that is, with chromosomes of gradually decreasing sizes. In order to measure the degree of asymmetry of a karyotype Petitpierre and Segarra (1985) have used the standard deviation (SD) of each chromosome relative length with respect to the averaged per cent length taken from the total complement length (TCL). They used this parameter but measuring the per cent of each chromosome length at mitotic metaphase with regard to the haploid TCL including the X but not the Y-chromosome, therefore, treating identically the species with or without a Y-chromosome. They calculated the SDs of asymmetry in 63 species and subspecies, whose karyotypes were mostly published, from the following ten genera of chrysomelines: the Holarctic *Timarcha* (Petitpierrre, 1970, 1976), and the Nearctics or Palaearctics *Calligrapha* (Robertson, 1966), *Chrysolina* (Petitpierre, 1983, 1999a, 1999b; Petitpierre and Segarra, 1985; Petitpierre *et al.*, 2004), *Oreina* (Petitpierre, 1999a), *Cyrtonus* (Petitpierre and Segarra, 1985; Petitpierre and Garneria, 2003), *Leptinotarsa* and *Labidomera* (Hsiao and Hsiao, 1983), *Phratora* (Petitpierre and Segarra, 1985), and the Neotropical *Araucanomela* (Petitpierre and Elgueta, 2006).

The increase in haploid chromosome number is generally associated with a decrease in asymmetry, the karyotypes are more symmetrical when they have more chromosomes, a clear trend which has also been reported in weevils (Curculionidae) by Lachowska *et al.* (1998). This does not mean at all an evident polarity towards increases in chromosome number by centric fissions, although it seems to be the more feasible trend in leaf beetles (Petitpierre and Segarra, 1985; Virkki, 1970, 1988; De Julio *et al.*, 2010). The chrysomelines support the reverse shifts in number by centric fusions: a) the origin of chiasmatic sex-chromosome systems neo-XY from the non-chiasmatic Xyp or XYp imply a translocation between an autosome and the X-chromosome, with the loss or fusion of the y-chromosome. The karyotype with the lowest number reported to date in chrysomelines, that of *Doryphora quadrisignata*, with 5 + neo XY meioformula (Vidal, 1984), has probably arisen by a centric fusion of this previous type plus several further fusions between autosomes. The meioformula of *Timarcha aurichalcea*, 8 + neoXY, the lowest one so far found in this genus, has been clearly demonstrated to be due to an all-arm translocation between a X-chromosome and one autosome bearing the rDNA loci, by fluorescent *in situ* hybridization (FISH) using a ribosomal DNA probe (Gómez-Zurita *et al.*, 2004), The origin of the strikingly asymmetric karyotype of *Chrysolina* (*Stichoptera*) *kuesteri* with 2n = 22 chromosomes (Petitpierre, 1983), can be presumably explained from a 24- chromosome species of the same subgenus, such as *Chrysolina latecincta*, because the largest autosome of the former has 21.30 per cent of the complement length while that of the latter has 16.46 per cent only, therefore, a centric fusion between this largest autosome and a smaller one of *Chrysolina latecincta*, or any other karyologically similar species of the subgenus *Stichoptera*, may have given rise after fixation to the largest autosome pair of *Chrysolina kuesteri* (Petitpierre, 1999b).

Gayane *et al.* (2012) studied the male karyotypes of *Acmaeodera pilosellae persica* Mannerheim, 1837 with 2n=20 (18+neoXY), *Sphenoptera scovitzii* Faldermann, 1835 (2n=38–46), *Dicerca aenea validiuscula* Semenov, 1895 – 2n=20 (18+Xyp) and *Sphaerobothris aghababiani* Volkovitsh et Kalashian, 1998 – 2n=16 (14+Xyp) using conventional staining and different chromosome banding techniques: C-banding, AgNOR-banding, as well as fluorochrome Chromomycin A3 (CMA3) and DAPI. It was shown that C-positive segments were weakly visible in all four species which indicate a small amount of constitutive heterochromatin (CH). There were no signals after DAPI staining and some positive signals were discovered using CMA3 staining demonstrating absence of AT-rich DNA and presence of GC-rich clusters of CH. Nucleolus organizing regions (NORs) were revealed using Ag-NOR technique; argentophilic material mostly coincides with positive signals obtained using CMA3 staining.

The rapid air drying technique with giemsa staining was used to study karyotype in insect and arachnids by James and Brown (1985). This method proved to be time saving for good results. Both meiotic and mitotic cells can be successfully treated with colchicines to increase the number of metaphase cells, although some polyploidy nuclei are sometime produced.

Since Lady bird beetles are biocontrol agents of a large number of insect pests, therefore, they should be attempted with respect to chromosomal biodiversity, C-banding and Bar-coding which will be helpful for correct identification of beetles.

The family scarabaeidae of Order Coleoptera is one of the widely studied Coleopteran families (Bisoi and Patnaik, 1991). Though phenotypically varied, most of its members (80), possess a common chromosomes formula, 9 AA + Xyp comprising chiefly biarmed chromosomes. Chromosome number, sexmachanism and male meiosis in seven scarabaeid beetles have been reported. Five of them *viz., Copris indicus, Orthophagus gazelle, O. hindu, Gymnopleurus gemmatus* and *Scarabaeius gangeticus* exhibit 2 n = 20, Xyp while, the remaining two, *Apogonia nigricans* and *A. ferruginea* have 2 n = 19, Xo. Structural rearrangement, loss of atosomes, loss of Y etc. have been contemplated for the evolution of karyotypes in the above species.

According to Dasgupta (1977) and Smith and Virkki (1978) about 235 species of this family are karyologically known. The diploid number ranges between 12 and 22 with a clear mode of 20 (9AA + Xyp). The sex-mechanisms encountered in this family Xyp, Xyr, Xy, XY, neoXY and XO of which Xyp is most common in above mentioned species of scarabaeids. The subfamily scarabaeinae is also chromosomally very heterogeneous (2n = 12 - 21) but the model number is 20, 18 Xyp. The subfamily melolonthinae is chromosomally more conservative as most of the species possess 2n = 20, Xyp. However, Bisoi and Patnaik (1991) are not very much certain about the presence of a nucleolus in all Xyp systems and they are in opinion that the Xyp system may be nucleolar in some and chiasmate in others.

Interest in cytogenetic studies of mosquitoes has grown considerably during the recent years and importance of the cytogenetic approach in the analysis of closely related species of *Anopheline* has been demonstrated by several workers. Karyotypes and polytene chromosome maps are now available for several anopheline species. However, there is paucity in cytogenetic knowledge of the tropical fauna, specially of oriental region (Pal *et al.*, 1981). According to Jayaprakash (1990) in terms of relative homologies in banding patterns, *Anopheles leucosphyrus* appear to *An. dirus*, *An. tessellatus* and *An. maculatus* while, *An. jeyporiensis* appears close to *An. dirus*, *An. tessellatus* and *An. aconitus*.

Comparative studies of the polytene chromosomes of mosquito species offer an excellent opportunity to understand their evolutionary and phylogenetic relationship. Closely related species and even sibling species may sometimes be distinguished on the basis of the banding patterns (Coluzzi and Sabatini, 1967). Morphologically similar species are expected to have more similar banding patterns than species with little morphological similarity.

According to Narang *et. al.* (1973) a band to band comparison between the chromosome map of *An. leucosphyrus* with those of some other closely related species of the subgenus *Cellia*, including *An. maculatus, An. tessellatus* (Narang *et. al.*, 1974) and *An. dirus* (Baimai *et. al.*, 1980) has been made which shows that *An. leucospyrus* shares more extensive homologies in the banding pattern with *An dirus* and *An. tessellatus* (Neomyzomyia - group) than *An. maculatus* (Neocellia - group). Similarly, a gross comparison of the chromosome map of *An. jeyporiensis* with those of some other closely related anopheline species of the subgenus *Cellia* has been reported in *An. tessellatus* (Narang *et al.*, 1974), *An. dirus* (Baimai *et al.*, 1980) and *An. aconitus* (Sharma *et al.*, 1980). *An. jeyporiensis* (Myzomyia - group) shared significant similarities in the banding pattern with *An. dirus* and *An. tessellatus* (Neomyzomyia - group) and *An. aconitus* (Myzomyia - group). Summarising the chromosomal picture *An. jeyporiensis, An. dirus, An. tesellatus* and *An. aconitus* represent steps differentiating these species in the same line of their relationship. Thus, banding pattern is undergone for a definite sequence of evolutionary changes within the subgenus *Cellia*. However, chromosomal comparison alone would be arbitary approach of evolution.

Cytological studies in recent times are being considered as one of the routes to the biotypic characterization of insect species. Hence, attempt has been made on Karyomorphological studies in Brown plants hooper *Nilaparvata lungens* (Stal.) (Hemiptera : Delphacidae) which is most destructive pest of rice in South and South - east Asia, China, Japan and Korea. Karyological studies of brown plant hopper culture of Annamalainagar revealed 2n = 30 chromosomes consisting of 14 pairs of autosomes in both sexes plus an unequal XY pair in males and supposed to be equal XX pairs in females confirming reports from other countries (Narayanasamy and Baskaran, 1990) of the forty odd species of the family Delphacidae, the chromosome numbers the chromosomal number reported ranged from 24 to 37 with peak at 29 (Bhattacharya and Manna, 1973). The presence of 30 chromosoes in BPH might suggest a case of frangmentation.

The aphids are tiny pest insects measuring hardly 2mm in body length and shows two bars or cornicles on dorsal of abdomen. Many of whom have altogather lost their sexual mode of reproduction (anholocyclic) secondarily, while, some others retain this mode for a brief period during the year (holocyclic) but reproduce parthenogenetically during the greater period. Hence, intra-specific Karyotypic variations have been studied by Kar and Rahman (1991) in two species namely, *Aphis gossypii* and

Toxoptera auranti as polyphagous species. Aphid belongs to order Hemiptera and family Aphididae. Blackman (1980) observed difference in the chromosomal morphology of some aphids within the species. According to Kar and Rahman (1991) all the different host plant samples of *A. gossypii* and *T. aurantii* had 2n = 8 chromosomes comprising one pair of 'long,' two pairs of 'medium' and one pair of 'short' chromosomes and lacking any primary constrictions. Although the diploid number was the same in each. There was some variations in their measurement values of individual pair of chromosomes. According to Kuznetsova (1980) the chromosomes of aphids are believed to be holokinetic in nature (White, 1973). Randomly occurring frangmentations and fusions altering the Karyotypes of some parthenogenetically reproducing aphids, some times producing structural heterozygotes (Blackman, 1971). For polyphagous species of aphids it is suggested that the 'standard' Karyotype' should have certain allowances for the naturally occurring variations within species and that a range of host - plant samples should be carefully compared for arriving at a reasonably accurate Karyotype. More over, host plant sample and places of collection should always be mentioned along with the described Karyotype. However, banding pattern plays vary crucial role in confirming species and host plant specific chromosomal changes needs to be exploited in near future. In general, chromosomal diversity in closely related species, will add great relevance in correct identification of the species and further biological, genetical and control of pest insects.

Cytogenetic studies that consider both the morphology and the chromosomal numbers of aphids can be extremely useful to the taxonomist (Blackman, 1981). According to Blackman (1981) the ubiquitous condition of holocentrism found in aphid chromosomes makes the proper recognition of each chromosome pair very difficult. In the holocentric chromosome the centromeric activity is diffused through the length of the chromosomes. Hence, Blackman (1980) says that due to such characteristics, aphid chromosomes are difficult material for the cytogeneticist. Because of the great variability in chromosomal number many workers believe the fact that the aphids have not yet achieved complete stability in their chromosomal number. Holocentric nature of chromosome is the main reason for unstability of chromosomal number of aphids.

de Celis *et al.* (1997) examined the chromosome set of the aphid species *Sitobion avenae, Schizaphis graminum* and *Methopolopium dirhodum* by means of conventional staining and C, NOR, *AluI* and Hae III banding methods. These species are important pests of several plants of economic importance in Brazil. In *S. avenae* there was no variation in the number of chromosomes while in *S. graminum* and *M. dirhodum* there was intra-specific variation. Interspecific differences in the response of banding treatments were observed. These techniques allowed the identification of several *S. graminum* chromosome pairs, only the AluI treatment was capable of inducing differential staining in *M. dirhodum* chromosomes and no clear patterns emerged when the *S. avenae* preparations were treated. The analysis of *S. graminum* chromosomes revealed a chromosome number of 2n = 8, as previously reported by Blackman and Eastop (1985). However, greater variation in such chromosome numbers was observed in the metaphase plates, after the screening of 25 individuals of this aphid, ranging from 2n = 6 to 2n = 8. Such variations are found in several insects appear to be common, but as stressed by White (1973), the most frequently occurring number in a group, can be considered the "type number". The differences in aphid chromosome numbers may be due to dissociation or fusions involving elements of the normal diploid set or to the presence of supernumerary B-chromosomes. Manicardi *et al.* (1991) observed a conspicuous banding pattern only in one of the higher chromosomes that lies in one intercalary band; particularly in X chromosomes, a single was indicated. The response of NOR banding treatment revealed only a single chromosome pair banded in *S. graminum*. Blackman and Eastop (1985) reported the chromosomal number of 2 n = 18, but for other species of the same genus viz. *M. festucae* and *M. frisicum* the 2 n = 16 chromosomal set was recorded by Blackman (1980). A higher variation of the karyotype within aphid species is expected than in other insects, due to their holocentrism. Thus, if the centromeric activity of these chromosomes is dispersed along its full length, broken chromosomal frangments are still capable of segregating at mytosis (Ris, 1942). However, Blackman (1980) is not supporting to Ris (1942). He say that there is no any strong evidence for stabilizing damaged chromosomes. In fact, the thelytokous reproduction of the aphids is a factor that allows karyotype variation within population of the same species.

Gomes *et al* (1997) studied karyotype evolution in wasps of the genus *Trypoxylon* (subgenus) *Trypargilum* (Hymenoptera : Sphecidae). The genus *Trypoxylon* is divided into two subgenera *Trypoxylon* and *Trypargilum*. The subgenus *Trypargilum* is confined to the western part of the Southern hemisphere, with a greater diversity of species in the neotropical regions (Amarante, 1991). The males of sub genus *Trypargilum* mate with females during construction of nest and remain present their as guard upto the completion of the nest. According to Coville (1982) all *Trypoxylon* species have their nest with spiders. Some species construct with mud while others use preexisting tubular cavities. Such cavities are further divided by the wasps into a linear series of cells with mud partition. *Trypargilum* wasps are solitary and the females of which construct their own nests.

From the genus *Trypoxylon* 12 species have been analyzed cytogetically (Gomes *et al*, 1997; Hoshiba and Imai, 1993). C-banding method was given by Sumner (1972), later, it was modified by Pompolo and Takahashi (1990). The highlights of C-banding method are given below -

Preparation of slides and treatments to slides after 3 days.

1. Hydrolyzation by immersion in 0.2 HCl for a variable period of time (3 to 5 min.)

2. Washing in distilled water for approximately 30s.

3. Immersion in 5 per cent $Ba(OH)_2$ in water bath at 60°C for 5 to 7 min.

4. Washing in 0.2 N HCl for about 30s and then in distilled water.

5. Immersion in 2 X SSC solution (Sodium Citrate and Sodium Chloride Saline), pH 7.0, in water bath at 60°C for 6-8 min and then in distilled water.

6. Staining with Giemsa (Merck) in 8 per cent 0.01 M phosphate butter, pH 6.8 and carefully washing in running water.

7. Some C-bands are obtained from slides submitted to standard staining.

8. Analyzation of slides and photography.

9. Slides are kept slanting on filter paper to drain the immersion oil. Then washed with battery of 3 flasks containing xylene, 5 min in each flask. The slides were then dried, destained by immersion in

0.2 N HCl and submitted to the procedure of C-banding. The slides are examined under light microscope using an immersion objective. Metaphases be selected for best quality of chromosomes.

Gomes *et al.* (1967) studied *Trypoxylon (Trypargilum) nitidum, T. (T.) lactilarse* and there unidentified species. The karyotype of *T. (T.) nitidum* (n = 15 and 2n = 30) consists of 10 pseudoacrocentric AM chromosomes (pairs 1, 6, 8, 10 and 15), 18 acrocentrics and two M^{cc} metacentrics (pair 14). The karyotype of the females of this species presents heteromorphism of pair 1, in which one of the chromosomes has a shorter long arm than the long arm of its homologue. Its C-bandiing pattern showed heterochromatin throughout the extension of one of the arms in the pseudoacrocentric chromosomes, on the short arm, in the pericentromeric region of the acrocentrics, and in the pericentromeric region of metacentric chromosomes. In the same species diploid karyotype formula consisted 2 K = 10 AM + 18 A + 2 M^{cc} and the diploid arm number was 2 AN = 42. While the same formula in *T. (T.) lactitarse* was 2 K = 28 A ^{MC} + 2 M^{cc} and the diploid arm number was 2 AN = 60. In *T. (T.) indium* and *T. (T.) lacitarse* the haploid chromosome number was same viz. n = 15 but differ in arm number (AN = 20, 21 and 30 respectively). However, morphological polymorphism observed in the chromosomes of pair 15 in *T. (T.) lactitarse* may be due to pericentric inversion in which metacentric chromosomes were converted to submetacentrics or vice versa, or may be due to a change in the functional centromere (Imai, 1991).

According to Imai *et al.* (1994) telocentric chromosomes originating from biarmed chromosomes by centric fission are converted to acrocentrics or pseudoacrocentrics by tandem growth of constitutive heterochromatin, and that this heterochromatin region may contain multiple dormant or inactive centromeres and telomeres that may be reactivated and change chromosome morphology. Hoshiba and Imai (1993) says that the number of chromosomes with a heterochromatic arm increases with increasing chromosome number and the correlation is a strong indication of the high telomeric instability of telocentric chromosomes in Hymenoptera.

Newton *et al.* (1974) studied X and Y chromosomes of *Aedes aegypti* (L.) distinguished by Giemsa C-Banding. The well known diploid karyotype of the mosquito, *Aedes aegypti* (L.) consists of three pairs of chromosomes since their correlation with the three genetic linkage groups are now referred to as 1, 3 and 2 in ascending order of size of these, chromosome 3 is submetacentric with a secondary constriction in the longer arm, whereas the largest and smallest are both metacentric. Bhalla (1973) has identified the shortest pair as the sex chromosomes but a reliable cytological basis for differentiating between the two has not yet been discovered. Since they are morphologically indistinguishable, as in other members of the tribe Culicini; Hence, for detecting the difference in internal structure

A Giemsa C-banding technique applied by Newton *et al.* (1974) to the mosquito *A. aegypti* has revealed a distinctive banding pattern which was described as a reliable means of distinguishing between the morphologically similar X and Y chromosomes during all stages of mitosis and meiosis. The difference noted was that the Y chromosome, unlike the X and the autosomes, was not c-banded in the centromere region. An intercalary band was also noted in one arm of X chromosomes and some Y chromosomes. According to Newton *et al* (1974) the distribution of these cytological markers throughout meiosis indicates that the sex locus somewhere within a pericentric region, the minimum extent of which includes both the intercalary band and the centromere. The data presented by Newton *et al.* (1974) are not inconsistent with the results of McDonald and Rai (1970) who, on the basis of radiation induced translocations, considered the sex locus to be approximately median in one arm.

Parise-Maltempi and Avancini (2001) studied C-banding and FISH in chromosomes of the blow flies *Chrysomya megacephala* and *Chrysomya putoria* (Diptera : Calliphoridae). Blow flies includes several common synanthropic forms, most of them with saprophagous habits. Some of blow flies are considered a serious public health problems as they cause myiasis in man and domestic animals. The blowflies have also tremendous importance in forensic science.

C. putoria has its origin from Africa and *C. megacephala* from Asia and Australia (James, 1970). Both the above species has been entered in Brazil in 1970 and now very common in Brazil. Morphological variations are found in the karyotypes of the species of family Calliphoridae, but chromosomal number is very stable at 2n = 12 with five autosomes and a heteromorphic sex pair (Boyes and Sheweel, 1975). According to Parise-Maltempi and Avancini (2001) *C. putoria* and *C. megacephala* contain 2n = 12 chromosomes, five metacentric pairs of atosomes and an large sex chromosome pair. There was no substantial difference in the karyotype morphology of these two species, except for the X-chromosome which was subtelocentric in *C. megacephala* and metacentric in *C. putoria* and was 1.4 times longer in *C. putoria*. All autosomes were characterized by the presence of a c-band in the pericentromeric region. *C. putoria* was also with interstitial band in pair III. The sex chromosomes of both species were heterochromatic, except for a small region at the end of the long arm of the X-chromosome. Above workers also detected Ribosomal genes in meiotic chromosomes by FISH and in both species the NOR was located on the sex chromosomes. On the basis of chromosomal features Parise-Maltempi and Avancini (2001) concluded that *C. putoria* species was introduced into Brazil in 1970s. *C. putoria* was morphologically very similar to *P. chloropyga* in Brazil. Based on the similarity of genitalia it was supposed to be the variant of *C. chloropyga*. However, Vllerich (1976) considered these two species to be distinct based on their cytogenetics.

According to Hadjiolov (1985), in many insects, including Diptera secondary constrictions and NORs are located together on heterochromatic sex chromosomes. Vllerich (1963) described the karyotype of *Lucilia cuprina* and showed that the nucleolus was associated with secondary constrictions present in the X and Y chromosomes.

Few chromosomal studies on the longhorned grasshoppers (Order : Orthoptera) are vailable from Turkey. According to Warchalowska - Sliwa (1998) in the family Tettigoniidae, the 2 n male chromosome number ranged from 20 to 35 and they exhibited a certain degree of conservation of karyotypes. In the sub-family Bradyporinae, 29 karyotypes of 25 species, distributed in the South Palearctic area have been described by Warchalowska - Sliwa (1998). In three species of *Zichyini* he characterized 2n male = 31 as a basic karyotype. However, Bradyporinae karyotypes ranged from 22 to 31 as XO or new XY sex determination mechanism. In the genus *Bradyporini* chromosome number ranged from 25-29. Similarly, in five species of the genus *Pycnogaster* the male chromosomal number was 29. Likely, in *Bradyporus macrogaster pancici* it was ranged from 25 to 29. In Orthoptera, the c-banding technique and Ag-staining of the nuoleolus organizer region (NOR) is normally used in comparative studies of populations, races and species. Paracentremeric c-bands were uniformly present in the long and medium sized chromosomes of *C. macrogaster*.

The sex determination XX female and XO male was found in C. *macrogaster*. In many insects chromosomal diversity is unknown and would be interesting aspects for their correct identity and utility in the sustainable development of the region.

Bibliography

Aiyar, T.V.R. (1924). An undiscribed Coccinellid beetle of economic importance. *J. Bombay Nat. Hist. Soc.* 30: **491-493**. (W.L. 25676).

Amanda, P.D.A., Cabral-De-Mello, D.C., Barros E Silva, A. E. and De Moura, R. D. C. (2009). Cytogenetic characterization of *Eurysternus caribaeus* (Coleoptera: Scarabaeidae): evidence of sex-autosome fusion and diploid number reduction prior to species dispersion. *Journal of Genetics*, 88: **177-182**.

Anderson, W. H. (1936). A comparative study of the labium of Coleopterous larvae. Smithson. *Misc. Coll.*, 95(13): **1-29**.

Angus, R.B., Wilson, C.J. and Mann, D. J. (2007). A chromosomal analysis of 15 species of Gymnopleurini; Scarabaeini and Coprini (Coleoptera; Scarabaeidae). *Tijdschr. Entomol.* 15: **201-211**.

Appels, R. and Peacock, W. J. (1978). The arrangement and evolution of highly repeated (satellite) DNA sequences with special reference to *Drosophila*. *Int. Rev. Cytol.* 8: **69-126**.

Arnett, R. H. (1963). The beetles of the United States. pp. **11- 12**. Washington.

Arrow, G. J. (1925). The fauna of British India, including Ceylon and Burma, Coleptera-Clavicornia, Erotylidae, Languridae and Emdomychidae, pp. **416** I Col. Pl. London.

Baimai V. (1988b). Population cytogenetics of the malaria vector *Anopheles lucosphyrus* group. *Southeast Asian J. Trop. Med. Pub. Hlth.* 19: **667-680**.

Balduf, W. V. (1935). The bionomics of entomophagous Coleoptera (13. Coccinellidae – lady beetles). pp. **220**. John S. Swift Co. Inc., Chicago, New York.

Banks, C. J. (1955). An ecological study of Coccinellidae associated with *Aphis fabae* Scop. on *Vicia faba*. *Bull. Ent. Res.* 46: **561-589**. (W.L. 10184).

Baratte S., Cobb M. and Peeters C. (2006a). Reproductive conflicts and multilation in queen less *Diacamma* atns. *Anim. Behav.* 72: **305-311**.

Baudry E., Peeters C., Brazier L. Veuille M. and Doums C. (2003). Shift in the behaviours regulating monogyny is associated with high genetic differentiation in the queenless ant *Diacamma ceylonense. Insects Soc.* 50: **390-397**.

Baverstock P. R., M. Gelder, A. Jahnke (1982). Cytogenetic studies of the Australian rodent, *Uromys caudimaculatus*, a species showing extensive heterochromatin variation. *Chromosoma* 84: **517-533**.

Benham, B.R., Muggelton, J. (1970). Studies on the ecology of *Coccinella undecimpunctata* Linn. (Coleoptera: Coccinellidae). *Entomologist*, **153-170**.

Benkevich, V. I. (1958). Biology of *Coccinella septempunctata* Uchem. *Zap. Orekh. Zuev Pedag. Inst.*, 11: **127-133**.

Berghella, L. and Dimitri, P. (1996). The heterochromatic rolled gene of *Drosophila melanogaster* is extensively polytenized and transcriptionally active in the salivary gland chromocenter. *Genetics*, 144: **117-125**.

Binaghi, G. (1941). Larvae Pupae di Chilocorini. *Memorie Soc. Ent. Ital.*, 20: **19-36**.

Bione, E. G., Camparoto, M. L. and Simoes Z. L. P. (2005a). A study of constitutive heterochromatin and nucleolus organizer regions of

Isocopris inhiata and *Dibroctis mimas* (Coleoptera, Scarabaeidae, Scarabaeinae) using C-banding, AgNO₃ staining and Fish techniques. *Genet. Mol. Biol.* 28: **111-116**.

Bione, E. G., Maura, R. C., Carvalho R. and Souza M. J. (2005b). Karyotype C- and fluorescence banding pattern, NOR location and Fish study of five Scarabaeidae (Coleoptera) species. *Genet. Mol. Biol.* 28: **376-381**.

Bonaccorsi, S. and Lohe, A. (1991). Fine mapping of satellite DNA sequences along the Y chromosome of *Drosophila melanogaster*; relationships between the satellite sequences and fertility factors. *Genetics*, 129: **177-189**.

Britten, R. J. and Kohne, D.E. (1968). Repeated sequences in DNA. *Science*, 161: **529-540**.

Brown, W. J. (1962). A revision of the forms of Coccinella L. Occurring in America north of Mexico (Coleoptera : Coccinellidae). *Can. Ent.* 94: **785-808**.

Brown, W. J., and R.de Ruette. (1962). An annotated list of the Hippodamiini of Northern America with a key to genera (Coleoptera : Coccinellidae). *Can. Ent.* 94: **643-652** (W.L. 13141).

Canepari, C. (1986). On some Coccinellids of Northern India and Nepal in the Geneva Museum of Natural History (Coleoptera, Coccinellidae). *Revue Suisse Zool.* 93: **21-36**.

Carbal-de-Mello, D. C., Oliveira, S. G., Ramos I. C. and Moura, R. C. (2008). Karyotype differentiation patterns in species of the subfamily Scarabaeinae (Scarabaeiae, Coleoptera). *Micron.* 38. **1243-1250**.

Carbal-de-Mello, D. C., Silva, F.A. B. and Moura, R. C. (2007). Karyotype characterization of *Eurysternus carybaeus*: The smallest diploid number among Scarabaeidae (Coleoptera, Scarabaeinae). *Micron.* 38. **323-350**.

Carson H. L., K. Y. Kaneshiro (1976). *Drosophila*i of Hawaii: Systematics and ecological genetics. Ann. Rev. Ecol. Syst. 7: **311-346**.

Casey, T. L. (1899). A revision of the American Coccinellidae. *J.N.Y. Ent. Soc.* 7: **71-169**.

Casey, T. L. (1908). Notes on the Coccinellidae. *Can. Ent.* 40: **349 - 421**.

Chapin, E.A. (1965). Coleptera: Coccinellidae. *Ins. Micronesia.*, 16: **189-254**.

Chapin, E.A. (1965). The genera of the Chilocorini (Coleoptera: Coccinellidae). *Bull. Mus. Comp. Zool.* 133: **227-271**.

Chapin, E.A. (1971). The Coccinellidae of Louisiana (Insecta : Coleoptera) Doctoral Dissertation, Louisiana State University.

Chapin, E.A. (1974). The Coccinellidae of Louisiana (Insecta : Coleoptera). *Agri. Exper. Sta. Doyle Chambers, Bull.* 682, pp. **1-87**.

Cherian P. T. (2000). On the status, origin and evolution of hot spots of biodiversity. *Zoo's Print.* IS (4): **247-251**.

Clayton, F. E. (1969). Variations in metaphase chromosome of Hawaiian Drosophilidae. *Univ. Texas Pub.* 6918: **95-110**.

Clayton, F. E. (1988). The role of heterochromatin in karyotype variation among Hawaiian pictured-wing *Drosophila. Pacific Science*, 42: **28-47**.

Cokendolphar J. C. and O. F. Francke. (1985). Karyotype of *Conomyrma flava* (Mc Cook) (Hymenoptera: Formicidae). *J. New York Entomol. Soc.* 92 (4): **349-351**.

Colamba M. S., Monteresino E., Vitturi R. and Zunino Z. (1996). Characterization of mitotic chromosomes of the scarab beetles *Glyphoderus sterquilinus* (Westwood) and *Bubos bison* (L.) (Coleoptera, Scarabaeidae) using conventional and banding technique. *Biol. Zentralbl.* 115: **58-70**.

Costa, C. (2000). Estado de conocimieento de los Coleoptera Neotropicales. In Hacia un proyecto CYTED para el inventarion y estimatcion de la diversidad entomologica en iberoamerica; prIBES-2000. Monografias tercer milenio. (ed. F. Martin-Piera, J.J. Morrone and A. Melic), 1: **99-114**.

Crotch, C. R. (1874). The revision of Coleopterous family Coccinellidae. pp. **311**., London (I, II).

Crotch, G. R. (1973). Revision of Coccinellidae of the United States. *Trans. Amer. Entomol. Soc.*, 4: **363-382**.

Crowson, R. A. (1960). The phylogeny of Coleoptera. *A. Rev. Ent.* 5(1), **111-134** (W.L. 3436).

Dange M. P., Ahish Rathod. (2010). Chomosome studies on four species of Scarabaeinae (Scarabaeinae: Coleoptera). *Natl. J. l. Sci.* 7 (2): 127-130.

Debach, Paul, (1964). Biological control of insect pests and weeds. PP. **1-843**. *Chapman and Hall Ltd. London.*

Dias, C.M., Schneider, M.C., Rosa, S.P., Costa, C. and Cella, D. M. (2007). The first cytogenetic report of fireflies (Coleoptera, Lampyridae) from Brazilian fauna. *Acta Zool.* 88: **587-589**.

Dover, G.A. (1982). Molecular drive: a cohesive mode of species evolution. *Nature*, 299: **111-116**.

Dover, G.A. and Flavell, R.B. (1984). Molecular co-evolution: DNA divergence and the maintenance of function. *Cell*, 38: **622-623**.

Duff, M. (1970). The chromosomes of four New Zealand insects. *New Zealand J. Sci.*, 13: **177-183**.

Dutrillaux, A. M., Xie, H. and Dutrillaux, B. (2007). High chromosomal polymorphism and heterozygosity in *Cyclocephala tridentate* from Guadalouoe; chromosome comparison with some other species of Dynastinae (Coleoptera: Scarabaeidae). *Cytogenet. Genome Res.* 119: **248-254**.

Dutt, G. R. (1927). Aphids and lady bird beetles. *Agric. India* 22(4). **291-292**.

Dyanechka, N. P. (1954). Coccinellids of the Ukrainian USSR. pp. **156**. Keiv (in Russia).

Ferreira, A. and Mesa, A. (2007). Cytogenetics studies in thirteen Brazilian species of Phaneropterinae (Orthroptera: Tettigonioidea; Tettigoniidae); main evolutive trends based on their karyological traits. *Neotrop. Entomol.* 36: **503-509**.

Fukumoto Y., Abe T. and Taki A. (1989). A novel form of colony organization in the "queenless" ant *Diacamma rugosum*. *Physiol. Ecol.*26: **55-61**.

Gage, J. H. (1920). Larvae of Coccinellidae. Illinois. *Biol. Mon.* 6(4): **232-294**.

Gatti, M. and Pimpinelli, S. (1992). Functional elements in *Drosophila melanogaster* heterochromatin. *Ann. Rev. Genet.* 26: **239-275**.

Ghani, M.A. (1962). A note on the identity of some species of genus *Ballia* (Coleoptera : Coccinellidae). *Proc. R. Ent. Soc. London (B)*, 31: **7-8**.

Gordon, R. D. and Chapin, E.A. (1983). A revision of the New World species of Stethorus Weise (Coleoptera : Coccinellidae). *Trans. Am. Ent. Soc.* 109: **229-276**.

Gordon, R.D. (1978). West Indian Coccinellidae II (Coleoptera) : some scale predators with keys to genera and species. *Coleopterists Bull.*, 32: **205-218**.

Gregory T. R., Nedved O., Adamovicz S. J. (2003). C-value estimates for 31 species of Lady bird beetles (Coleoptera: Coccinellidae). *Hereditas.* 139: **121-127**.

Gronenberg W. and Peeters C. (1993). Central projections of the sensory hairs on the gemma of theant *Diacamma*: substrate for behavioral modulation? *Cell Tissue Res.*273: **401-415**.

Hagen, K.S. (1966). Laboratory studies on the reproduction of *Adalia bipunctata* (Coleoptera : Coccinellidae). *Entomologia Exp. Appl.* 9: **200-204**.

Halffter, G. and Favila, M. E. (1993). The Scarabaeinae (Insecta: Coleoptera) an animal group for analyzing, inventorying and monitoring biodiversity in tropical rainforest and modified landscapes. *Biol. Int.* 27: **15-21**.

Heitz, E. (1928). Das heterochromatin der Moose. *Jahrb. Wiss.* Bot. 69: **762-818**.

Hewitt, G.M. (1979). Orthoptera; grasshoppers and crickets. Gerbruder Borntrager, Berlin.

Holldobler B. and Wilson E. O. (ed.) (1990). The Ants. Harvard University Press, Cambridge, USA.

Imai H. T., RH Crozier., RW Taylor. (1977). Karyotype evolution in Australian ants. *Chromosoma* 59: **341-393**.

Irick, H. (1994). A new function of heterochromatin. *Chromosoma*, 103: **1- 3**.

John B. (1981). Heterochromatin variation in natural populations. *Chromosome Today* 7: **128-137**.

John B., Lewis K. R. (1960). Neucleolar controlled segregation of the sex chromosomes in beetles. *Heredity*. 15: **431-439**.

John, B. (1981). Heterochromatin variation in natural populations. *Chromosome Today*, 7: **128-137**.

John, B. and Miklos, G.L.G. (1979). Functional aspects of satellite DNA and heterochromatin. *Int. Rev. Cytol*. 58: **1-114**.

Joseph T. M., Biju A. and Shekha V. (2002). Environmental importance of *Sacred groves* and their conservation. *Millennium Zoology*, 2 (1): **22-25**.

Juan C., Pons J., Petitpierre E. (1993). Localization of tandamly repeated DNA sequences in beetle chromosomes by fluorescent *in-situ* hybridization. *Chromosome Res*. 1: **167-174**.

Kacker, R. K. (1970). Studies on chromosomes of Indian Coleoptera. IV; In nine species of family Scarabaeidae. *Nucleus*, 13: **126-131**.

Kapur, A. P. (1947). A revision of the tribe Aspidimerini Weise (Coleoptera: Coccinellidae).

Kapur, A. P. (1948a). On the old World species of the genus *Stethorus* Weise (Coleoptera: Coccinellidae). *Bull. Ent. Res*. 39: **297-320**.

Kapur, A. P. (1948b). On the Indian species of *Rodolia* Mulsant (Coleoptera: Coccinellidae). *Bull. Ent. Res*. 39: **531-538**.

Kapur, A. P. (1950). The biology and external morphology of the larvae of Epilacninae. *Bull. Ent. Res*. 41: **161-208**.

Kapur, A. P. (1954). Systematic and biological note on the San Jose scale in Kashmir with description of a new species (Coleoptera: Coccinellidae). *Rec. Indian Mus.*, 52: **257-274**.

Kapur, A. P. (1955). Coccinellidae of Nepal. *Rec. Indian Mus.*, 53: **309-338**.

Kapur, A. P. (1963). The Coccinellidae of the third Mount Everest Expedition, 1924 (Coleoptera). *Bull. Brit. Mus. (Nat. Hist.), Ent*. 14: **1-48**.

Kapur, A. P. (1967).The Coccinellidae (Coleoptera) of the Andamans. *Proc. Nat. Inst. Sci. India*. 32(B), **148-189**.

Kapur, A. P. (1969). On some Coccinellidae of the tribe Telsimini with description of new species from India. *Bull. Syst. Zool. Calcutta,* 1(2): **45-56.**

Kapur, A. P. (1972). The occinellidae (Coleoptera) of Goa. *Rec. Zool. Survey India,* 62: **309-320.**

Karagyan, G., Kuznetsova, V.G. and Lachowska, D. (2004). New cytogenetic data on Armenian buprestids (Coleoptera, Buprestidae) with a discussion of karyotype variation within the family. *Folia. Biol.* 52: **151-158.**

Katzman M. T. and G. Cale (1990). Tropical forest preservation using economic incentives: *Bioscience,* 40: **827-832.**

King M. (1993). Species evolution: The role of chromosome change. Cambridge University Press Cambridge, UK.

Kitzmillar, J. B. (1976). Genetics, cytogenetics and Evolution of mosquitoes. *Adv. Genet.* 18: **316-433.**

Korschefsky, R. (1931). Coccinellidae 1. Coleoptorum Catalogus, 23(118) Den Haag. W. Junk. **224** pp.

Korschefsky, R. (1931. Coccinellidae I. PP **224**. Coleopterorum Catalogues Pars 118. Berlin.

Korschefsky, R. (1932). Coleoptererorum Catalogues Pars **118**. Coccinellisae I. pp. 224, Berlin.

Krebs C. J. (1989). Ecological methodology. Harper and Raw publishers New York.

Lahiri, M. and Manna, G. K. (1969). Chromosome complement and meiosis in nine species of Coleoptera. *Proc. 56th Ind. Sci. Cong. Pt.,* 3: **448-449.**

Le, M.H. and Duricka, D. (1995). Islands of complex DNA are widespread in *Drosophila* centric heterochromatin. *Genetics,* 141: **283-303.**

Lewis, G. (1873). Notes on Japanese Coccinellidae. *Ent. Mon. Maq.,* 10(2): **54-56.**

Lewis, G. (1879). On certain new species of Coleoptera from Japan. *Ann. Maq. Nat. Hist. Lond.* (5)4: **459-467.**

Lewis, G. (1896). On the Coccinellidae of Japan. *Ann. Maq. Nat. Hist. Lond.* (6) 17: **22-41.**

Macaisne, N., Dutrillaux, A.M. and Dutrillaux, B. (2006). Meiotic behavior of a new complex X-Y autosome translocation and amplified heterochromatin in *Jumnos ruckeri* (Saunders) (Coleoptera, Scarabaeidae), Cetoniinae). *Chrom. Res.* 14: **909-918.**

Machado, V., Galian, J., Araujo, A. M. and Valente, V. L. S. (2001). Cytogenetics of eight neotropical species of *Chouliognathus* Henzt, 1830; implications on the ancestral karyotype in Cantharidae (Coleoptera). *Hereditas*, 134: **121-124.**

Martins, V. G. (1994). The chromosome of five species of Scarabaeidae (Polyphaga, Coleoptera). *Naturallia*, 19: **89-96.**

Mayne, W.W. (1953). *Cryptolaemus montrouzieri* Mulsant in South India. *Nature*, **172: 185.**

Mayr E. (1963). Animal species and evolution. Cambridge, mass: Harvard Univ. Pr.

Miyatake, M. (1961a). The East-Asian Coccinellid-beetles preserved in the California Academy of Sciences, tribe Hyperaspini. *Mem. Ehime Univ.* (6) 6: **67-86.**

Miyatake, M. (1961b). The East-Asian Coccinellid-beetles preserved in the California Academy of Sciences, tribe Serangiini- *Mem. Ehime Univ.* (6) 6: **135-146.**

Miyatake, M. (1961c). The East-Asian Coccinellid-beetles preserved in the California Academy of Sciences, tribe Hyperaspini. *Mem. Ehime Univ.* (6) 6: **147-155.**

Miyatake, M. (1967). Notes on some Coccinellidae from Nepal and Darjeeling District of India (Coleoptera). *Trans. Shikoku Ent. Soc.* 9(3): **69-78.**

Miyatake, M. (1970). The East-Asian Coccinellid-beetles preserved in the California Academy of Sciences, tribe Chilocorini. *Mem. Ehime Univ.* 14(2): **303-340.**

Miyatake, M. (1985). Coccinellidae collected by Hokkaido University expedition to Nepal, Himalaya, 1968 (Coleoptera). *Insecta Matsumarana*, 30: **1-33.**

Mola, L.M., Papeschi, A.G. and Taboada, C. (1999). Cytogenetics of seven species of dragonflies. *Hereditas*, 131: **147-153**.

Moura, R.C., Sounza. M. J., Melo, N.F. and Lira-Neto, A.C. (2003). Karyotypic characterization of representatives from Melolonthinae (Coleoptera: Scarabaeidae): Karyotypic analysis, banding and fluorescent *in situ* hybridization (FISH). *Hereditas*, 138: **200-206**.

Mulsant, M. E. (1866). Monographic des Coccinellides Ire partie Coccinelliens. pp. **1-294**, Paris.

Myers N. (1988). *The environmentalists* 8: **187-208**.

Nutan Karnik, Channaveerappa, H., Ranganath, H. A. and Gadagkar, R. (2010). Karyotype instability in the ponerine ant genus *Diacamma*. *Journal of Genetics*, 89: **173-182**.

Pajni, H. R. and Singh, J. (1982). A report on the family Coccinellidae of Chandigarh and its surrounding areas (Coleoptera). *Res. Bull. Punjab Univ. Sci.* 33: **79-86**.

Pajni, H. R. and Verma, S. (1985). Studies on the structure of the male genetilia in some Indian Coccinellidae (Coleoptera). *Res. Bull. Punjab Univ. Sci.* 36: **195-201**.

Pardue, M.L. and Henning, W. (1990). Heterochromatin: junk or collector's item. *Chromosome*, 100: **3-7**.

Pathak S., TC Hsu, FE Arrighi. (1973). Chromosomes of *Peromyscus* (Rodentia, Cricetidae). IV. The role of heterochromatin in karyotype evolution. *Cytogenet. Cell Genet.* 12: **315-326**.

Patil, V. J. and Sathe, T. V. (2003). Predators and pest management. pp. **1-174**. Daya Publishing House New Delhi.

Pattarudraiah, M. and Channabasavanna, G. P. (1953). *Beneficial Coccinellids of Mysore – I. Indian J. Ent. 15:* **87-96**. *(W.L. 22997)*.

Pattarudraiah, M. and Channabasavanna, G. P. (1955). Beneficial Coccinellids of Mysore – II. *Indian J. Ent.* 17: **1-5**. (W.L. 22997).

Pattarudraiah, M. and Channabasavanna, G. P. (1956). Some beneficial Coccinellids of Mysore – III. *J. Bombay Nat. Hist. Soc.*, 54: **156-159**.

Patton J. L, SW Sherwood. (1982). Genome evolution in pocket gophers (genus *Thomomys*). 1. Heterochromatin variation and speciation potential. *Chomosoma* 85: **149-162**.

Peacock, W. J., Dennis, E.S., Rhoades, M.M. and Pryor, A.J. (1981). Highly repeated DNA sequences limited to knob heterochromatin. *Proc. Natl. Acad. Sci. USA.* 78: **4490 - 4494**.

Peacock, W.J., Lohe, A.R., Gerlach, W.L. Dunsmuir, P., Dennis, E.S. and Appels, R. (1977). Fine structure of and evolution of DNA in heterochromatin. *Cold Spring Harbor Symp. Quant. Biol.* 42: **1121-1135**.

Peeters C. (1993). Monogyny and polygyny in ponerine ants with or without queens. In *Queen number and sociality in insects* (ed. L. Keller), pp. **234-261**. Oxford University Press, UK

Peeters C. and Billen J. (1991). A novel exocrine gland inside the thoracic appandages ("gemmae") of the queenless and *Diacamma australe*. *Experientia.* 47: **229-231**.

Peeters C. and Higashi S. (1989). Reproductive dominance controlled by multilation in the queenless ant *Diacamma Australe. Naturwissenschaften.* 76: **177-180**.

Peeters C., Billen J. and Holldobler B. (1992). Alternative dominance mechanisms regulating monogyny in the queenless ant genus *Diacamma. Naturwissenschaften.* 79: **572-573**.

Petitpierre, E. and Garneria, I. (2003). A cytogenetic study of the leaf beetle genus *Cyrtonus* (Coleoptera, Chrysomelidae). *Genetica,* 119: **193-199**.

Petitpierre, P. Kippenberg, H. Mikhailov, Y. and Bourdonne, J. C. (2004). Karyology and Cytotaxonomy of the Genus *Chrysolina* Motschulsky (Coleoptera, Chrysomelidae). *Zool. Anz.* 242: **347-352**.

Poggio, M. G., Bressa, M. J. and Papeschi, A.G. (2007). Karyotype evolution in Reduviidae (Insecta: Heteroptera) with special reference to Stenopodainae and Harpactorinae. *Comp. Cytoge.* 1: **159-168**.

Ramaswamy K., Peeters C. Yuvana SP., Varghese T., Pradeep H. D., Dietemann V. *et al.* (2004). Social multilation in the ponerine ant *Diacamma*: cues originate in the victims. *Insects Soc.* 51: **410-413**.

Ranganath, H.A. and Hegade, K. (1982). The chromosomes of two Drosophila races; *D. nasuta* and *D. albomicana. Chromosoma,* 85: **83-92**.

Ray-Chaudhari S. P. and Pyne C. K. (1954). Time intensity factor in the production of dicentric bridges with Gamma rays of radiation during meiosis in the grass hopper, Gesonula punctifrons, Science, N. Y. 119: **685-686**.

Rojers W. A. (1991). In: Ecology and sustainable development (ed. Gopal). pp. **81-83**

Rozec, M. (1994). A new chromosome preparation technique for Coleoptera (Insecta). *Chromosome Res*. 2: **76-78**.

Saha, A. K. (1973). Chromosomal studies of the Indian Coleopterans (Indian Beetles). *Cytologia*, 38: **363-373**.

Sasaji, H. (1968a). Phylogeny of the family Coccinellidae (Coleptera). *Etizenia*. 35: **1-37**.

Sasaji, H. (1968b). Description of the Coccinellid larvae of Japan and the Ryukyus (Coleoptera). *Mem. Fac. Edn. Fukui Univ.* 2, (18): **93-136**.

Sasaji, H. (1971). Fauna Japonica Coccinellidae (Coleoptera). pp. **340**. Academic Press, Japan.

Sathe, T. V. (1986). Biology of *Cotesia diurnii* R and N (Hym. : Braconidae) a larval parasitoid of *Exelastis atomosa* Walsingham. *Oikoassay*, 3: **31-33**.

Sathe, T. V. (2004). Vermiculture and organic farming. pp. **1-122**. Daya Publishing House New Delhi.

Sathe, T. V. and Bhosale, Y. A. (2001). Insect pest predators. pp. 1-167. *Daya Publishing House New Delhi*.

Shaarawi, F. A. and Angus, R.B. (1991). A chromosomal investigation of five European species of Anacaena Thomson (Coleoptera: Hydrophilidae). *Entomol. Scand.* 21: **415-426**.

Sicard, A. (1907). Coleopteres Coccinellides du Japan. *Bull. Mus. Hist. Nat. Paris.,* 1907: **211**.

Sicard, A., (1909). Revision des Coccinellides de la faune. *Ann. Soc. Ent. Fr.,* 78: **68-165**.

Simmonds, F. J. (1963). Genetics and biological control problems. *Entomophaga,* 17: **251-264**.

Smith S. G. (1950). The cyto-taxonomy of Coleoptera. *Canad. Etomol.* 82: **58-68**.

Smith, S. G. and Virkki, N. (1978). Coleoptera, p. 59-70. In B. John (ed.), Animal cytogenetics. Berlin, Germany, Gebruder Borntraeger, **366** p.

Subbarao S. K, K. Vasantha, T. Adak, VP. Sharma. (1983). *Anopheles culicifacies* complex: Evidence for a new sibling species, species C. *Ann. Ent. Soc. Am.* 76: **985-988**.

Subramaniam, T. V. (1923). Some coccinellids of South India. Pp. **108-118**. Rep. Proc. Fifth Ent. Meeting, Pusa.

Takenouchi, Y. (1972). A note on the cytology of *Zabrotes subfasciatus* Boh (Coleoptera: Bruchidade). *Jap. J. Genet.* 47: **69-70**.

Timberlake, P. H. (1943). The Coccinellidae or lady beetles of the Koebele Collection part II, Hawaii. *Plant Rec.,* 47: **1-67**.

Tulloch G. S. (1934). Vestigial wings in Diacamma. *Ann. Ent. Soc. Am.* 27: **273-277**.

Veuille M., Brusadelle A. Brazier L. and Peeters C. (1999). Phylogenetic study of a behaviourial trait regulating reproduction in the ponerine ant *Diacamma.* in social insects at the turn of the millennium (ed. M. Schwarz and K. Hogendoorn), pp. **442**. 13th Congress of the International Union for the Study of Social Insects (IUSSI), Adelaide, Australia.

Virkki, N. (1957). Structure of the testis follicle in relation to evolution in Scarabaeidae (Coleoptera). *Can. J. Z.,* 35: **265-277**.

Vitturi, R.; Colomba, M.; Volpe, N; Lannino, A. and Zunino, M. (2003). Evidence for male XO sex-chromosome system in Pentdon bidins Punctatum (Coleoptera, Scarabaeoidea, Scarabaeidae) with X linked 18S – 28S rDNA clusters. Genes. Genet. Syst., 78: **427-432**.

Wakahama, K.I., Shinohara, T., Hatsumi, M., Uchida, S. and Kitagawa, O. (1983). Metaphase chromosome configuration of the *immigrans* species group of Drosophila. *Jpn. J. Genet.* 58: **315-326**.

Ward, B. L. and Heed, W.B. (1970). Chromosome phylogeny of *Drosophila pachea* and related species. *J. Hered.* 61: **248-258**.

Watson, W. Y. (1956). A study of the phylogeny of the genera of the tribe Coccinellini (Coleoptera). *Centr. Roy. Ont. Mus. Toronto (Zool.),* 42: **1-52.**

Weimarck A. (1975). Heterochromatin polymorphism in rye karyotype as detected by the giemsa C-banding technique. *Hereditas.* 79: **293-300.**

Wheeler W. M. (1915). On the presence and absence of cocoons among ants, the nest-spinning habits of the larvae and the significance of the black cocoons among certain Australian species. *Ann. Ent. Soc. Am.* 8: **323-342.**

White M. J. D. (1973). Animal cytology and evolution. 3rd Ed. London: Cambridge Univ. Pr.

White, M. J. D. (1978). Modes of speciation. San Francisco: WH Freeman.

Wilson, C. J. and Angus, R.B. (2004). A chromosomal investigation of seven Geotrupids and two Aegialiines (Coleoptera, Scarabaeoidea). *Nouv. Rev. Entomol.* (N.S.). 21: **157-170.**

Wilson, E. O. (1971). The insect societies. The Belknap Pess of Harvard University Press, Cambridge, USA.

Wilson, F. D., Wheeler, M. R., Harget, M., and Kambysellis, M. (1969). Cytogenetic relations in the *Drosophila nasuta* subgroup of the *immigrans* group of species. *Univ. Texas Publ.* 6918: **207-253.**

Yadav, J. S. and Pillai, R. K. (1979). Evolution of karyotype and phylogenetic relationship in Scarabaeidae (Coleoptera). *Zool. Anz. Jena.* 202: **105-118.**

Yoon, J. S. and Richardson, R. H. (1978). Evolution in Hawaii Drosophilidae. III. The microchromosome and heterochromatin of *Drosophila. Evolution,* 32: **475-484.**

Zuckerkandi, E. and Henning, W. (1995). Tracking heterochromatin. *Chromosoma,* 104: **75-83.**

Index

A

Acmaeodera pilosellae persica
   125, 128

Allium porrum 47

Allium sativum 47

Anopheles aconitus 136

An. dirus 136

An. jeyporiensis 136

An. leucosphyrus 136

An. leucospyrus 137

An. maculatus 136

An. tessellatus 136

Aphis. craccivora 81, 83, 86

Aphis gossypii 137

Aphodius fimetarius 129

Apogonia nigricans 135

Arachis hypogaea 58

Ascaris megalocephala 46

Autoserica assamensis 120

C

Calotropis gignatea 58

Carthumus tinctorius 58

Chalcophora mariana 125

Chironomids 40

Chironomous thumii thumii 30,47

C.t. piger 30

Chrysolina 129

Chrysolina carnifex 133

Chrysolina diluta 133

Chrysolina fuliginosa 133

Chrysolina gypsophilae 133

Chrysolina haemoptera 133

Chrysolina helopioides 133

Chrysolina interstincta 133

Chrysolina lepida 133

Chrysolina rufoaenea 130

Coccidula 76, 114

Coccinella 88

Coccinella cincta 86

Coccinella discolor 87

Coccinella septempunctata 89

Coccinella tabacI 89

Coccinella transversalis 90

Copris fricator 112

Copris indicus 135

Coreid plant bugs 39

Culex quinquepunctata 89

C. transversalis 89

C. trifasciata perflexa 89

Cyrtonus 129

D

Desmogramma 131

Diacamma 121

Dicerca aenea validiuscula 126, 127, 128

Diprion 49

Diprion simil 49

Doryphora quadrisignata 129, 134

Drosophila melanogaster 112

D. birchii 112

D. ceylonensei 121

D. disjuncta 112

D. kikkawai 112

D. melanogaster 113

D. serido 112

E

Epilachna vigintioctopunctata 116, 117, 118

Euchroma gigantea 127

Eurysternus caribaeus 111, 119

F

Foenicaluns vulgare 86

G

Gaillardia bulchella 58

Geotrupes mutator 115

Glycine soja 58

Gossypium hirsutum 81, 84

Grasshoppers 38

Gymnopleurus dejeani 110, 111

Gymnopleurus gemmatus 111,135

H

H. convergens 88

H. indica 88

H. variegata 88

Harmonia maharashtri 94, 96, 104.

H. soyabiniI 84

Helianthus annus 58

Heliocopris bucephales 110

Hippodamia 88

I

Illeis. bistigmosa 86

I. gavari 86

I. therioaphis 86

L

Leptinotarsa behrensi 133

Leptinotarsa decemlineata 133

Leptinotarsa lineolata 133

Leucaena retusa 58

Leucocera 131

M

Maistor sp. 40

Menochiius patilensis 94,98,106.

Menochilus sexmaculatus 91, 93, 102, 116

Menochilus aphidivouri 94, 105

Menochilus marathi 81, 100, 108, 109

Menochilus palusi 82, 94, 99, 107

Menochilus sangliensis 78, 94, 95, 103

Menochilus sathei 82

Menochilus vaishali 81, 91, 92, 101

Myzus persicae 90

N

Neodiprion 49

Nicotiana 41

N. tabacum 90

Nilaparvata lungens 137

O

Ophioglossum reticulatum 46

Orthophagus gazelle 135

O. hindu 135

P

Palembus dermestoides 124

Phyllophaga 124

Phyllophaga (Phytalus) vestita 124

Platyphora 131

R

Rhophalosiphum maidis 79, 81, 87, 88

Rh. litura 76

Rhyzobius 76, 114, 115

Rh. chrysomeloides 114

S

Scarabaeius gangeticus 135

Scarabaeus pedellus 129

Sciarids 39

Sorghum vulgare 79, 81.

Sphaerobothris aghababiani 126, 127, 128

Sphenoptera mesopotamica 128

Sphenoptera scovitzii 127

Strichosa 131

T

Timarcha 129, 134

Timarcha pimelioides 133

Toxoptera auranti 138

Typhaeus typhoeus 115

V

V. cardoni 87

V. kiotoensis 87

V. vulgeri 87

Vernia discolor 87

Vernia indica 87

Vernia polyphagae 87

Vigna unguiculata 58

Y

Yucca arkansana 47

Z

Zabrotes subfasciatus 122, 124

Zea mays 56, 81, 84, 88

Zygogramma 131